Measuring Distances in Africa

Applying the Spherical Law of Cosines for Sides

AF887747

Copyright @ 2018 Dale Johanson
All rights reserved.
Curve Beneath the Cloud LLC
ISBN-13: 978-0-9978303-1-6

Contents

Preface

Spherical trigonometry is a truly fascinating subject that declined in popularity in the middle of the twentieth century and is rarely taught today. In this activity book, you will use one of the laws from that subject to calculate the approximate distance between two capital cities in Africa.

In chapter one, "Laying the Foundation", you will find a short list of items from plane trigonometry that you may consider reviewing. You will also find a proof of the spherical law of cosines for sides which you can either read or skip. The chapter then continues with a more in depth look at the first example of chapter two and concludes with information and illustrations to help you create similar illustrations in the problems you will be solving.

Chapter two has a few examples followed by problems where you can show your work in calculating the distances between capitals in Africa. After you have finished your calculations on a problem, you can check the answer in the solutions found in chapter three. In the index of cities and countries, you will find a listing of the page numbers where a problem and solution involving a particular capital or country can be found. In some cases, more than one capital may exist. For example, a country may have a legislative capital as well as an administrative capital. If that is indeed the case, I will reference the latitude and longitude of the administrative capital. There are a total of three examples and 24 problems with solutions in this activity book. I hope you enjoy calculating the distances between capitals in Africa using the spherical law of cosines for sides!

Dale Johanson
April 11, 2018

Chapter 1

Laying the Foundation

Before you embark on your journey around the world via the spherical law of cosines for sides you may want to study and/or review a few key items from plane trigonometry. The main items include the

1. sine function
2. cosine function
3. inverse cosine function
4. arc length formula
5. Pythagorean Theorem
6. law of cosines.

In addition to those items from plane trigonometry, other items of importance that you may want to study and/or review include

1. the equator
2. the prime meridian
3. latitude
4. longitude

In the problems throughout the book, a longitude will either be referenced in degrees west of the prime meridian or in degrees east of the prime meridian.

You can use either a scientific or graphing calculator. Make sure the calculator is in degree mode, and float mode. Basically, we don't want the calculator to round to a fixed number of decimal places.

After you've completed the tasks outlined above you can either read through the proof of the spherical law of cosines for sides which starts on the next page, or you can skip to page 13 where all the steps involved in calculating the approximate distance between two capitals are explained.

Imagine a plane intersecting a sphere at single point A. Extend a line segment from the center of the sphere, A_1 , outward intersecting the sphere at point B and then intersecting the plane at point B_1. Next, extend another line segment from the point, A_1 , outward intersecting the sphere at point C and then intersecting the plane at point C_1. Finally, extend another line segment from the point, A_1 , outward intersecting both the sphere and the plane at point A. Unlike the sides of the triangle in a plane, the sides of the spherical triangle formed by the arcs a, b, and c, respectively, are themselves angles. The given sides a, b, and c also represent arcs on great circles. A great circle is the circle formed by the intersection of a plane passing through the center of a sphere and the sphere itself.

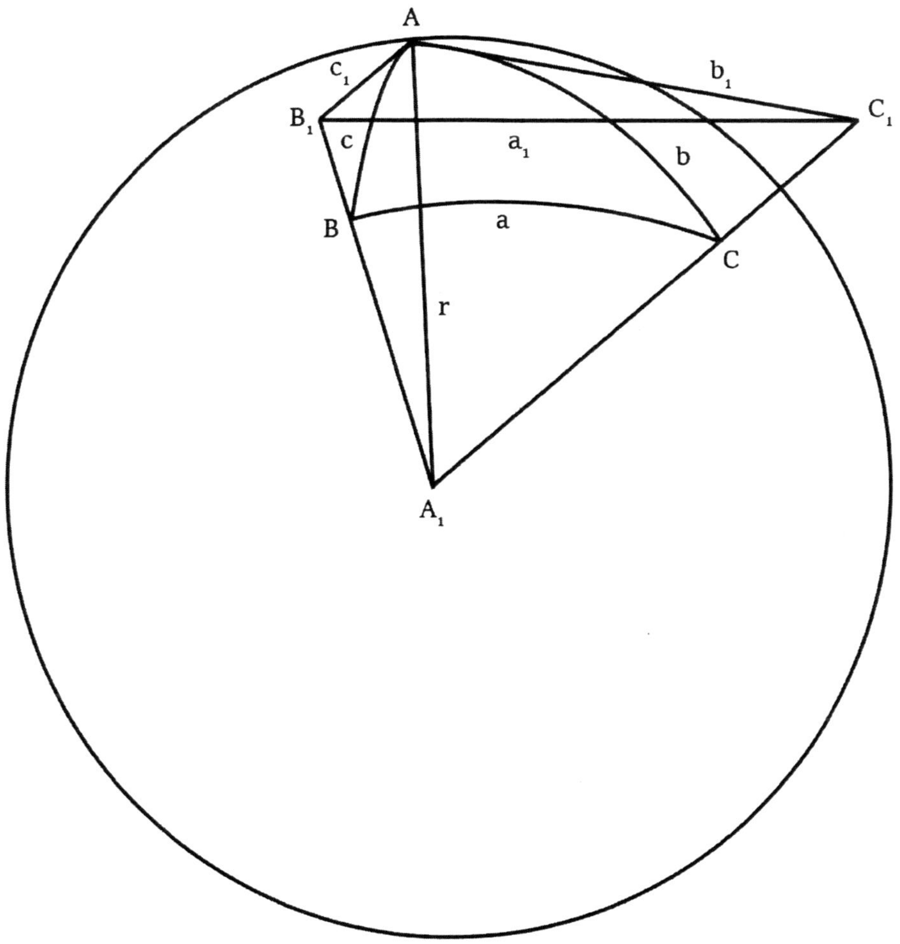

Figure # 1

As we proceed through the steps, we consider four triangles created in Figure # 1.

We first examine the triangle,

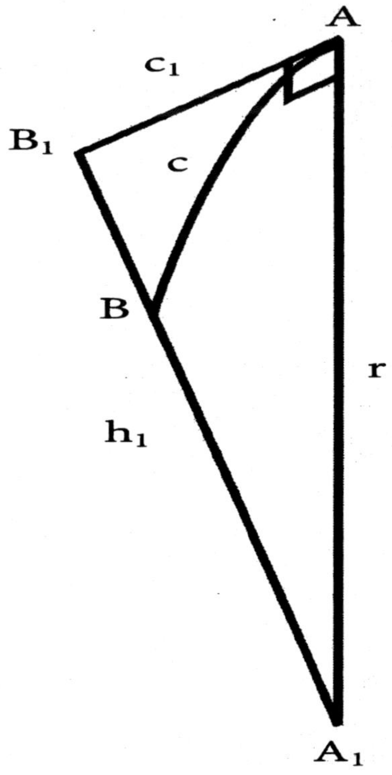

Figure # 2

Let h_1 represent the hypotenuse of the right triangle above. It follows that,

$\cos c = \frac{r}{h_1} \Rightarrow h_1 \cos c = r \Rightarrow h_1 = \frac{r}{\cos c}$ and $\sin c = \frac{c_1}{h_1} \Rightarrow h_1 \sin c = c_1 \Rightarrow c_1 = \left(\frac{r}{\cos c}\right) \sin c$

$\Rightarrow c_1 = \frac{r \sin c}{\cos c}$ (equation # 1)

Second, we examine the triangle,

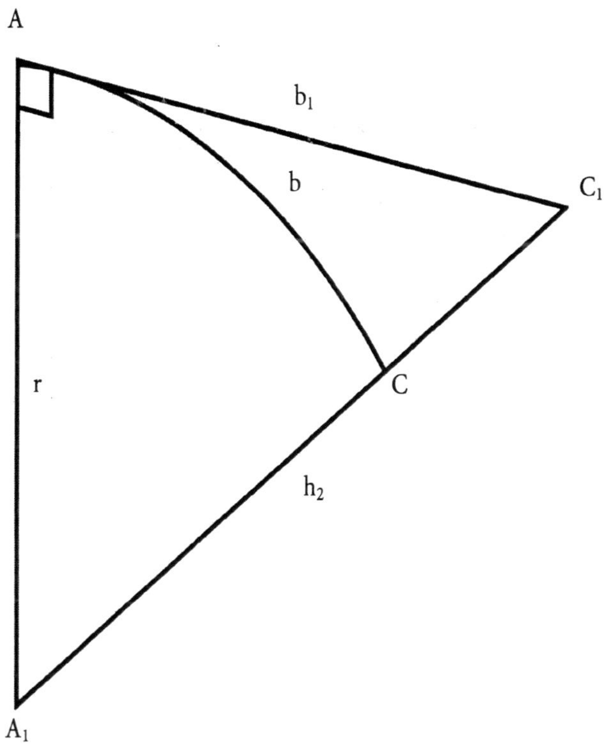

Figure # 3

Let h_2 represent the hypotenuse of the right triangle above. It follows that,

$\cos b = \frac{r}{h_2} \Rightarrow h_2 \cos b = r \Rightarrow h_2 = \frac{r}{\cos b}$ and $\sin b = \frac{b_1}{h_2} \Rightarrow h_2 \sin b = b_1 \Rightarrow b_1 = \left(\frac{r}{\cos b}\right) \sin b$

$\Rightarrow b_1 = \frac{r \sin b}{\cos b}$ (equation # 2)

Third, we consider the triangle on the plane.

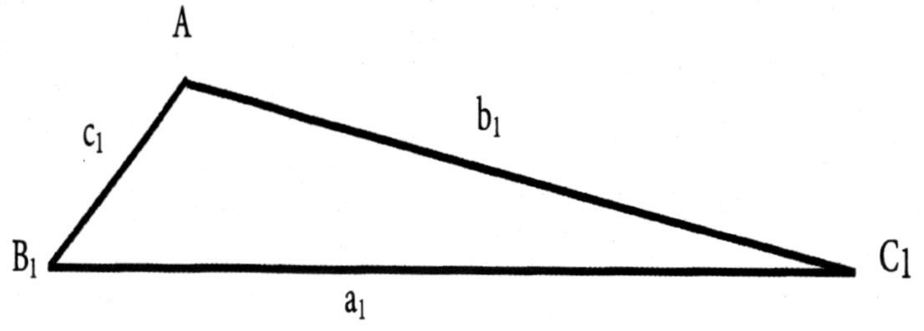

Figure # 4

Using the law of cosines on a plane,

$$a_1{}^2 = b_1{}^2 + c_1{}^2 - 2b_1c_1 \cos A$$

$$= \left(\frac{r\sin b}{\cos b}\right)^2 + \left(\frac{r\sin c}{\cos c}\right)^2 - 2\left(\frac{r\sin b}{\cos b}\right)\left(\frac{r\sin c}{\cos c}\right)\cos A \quad \text{using equations \# 1 and \# 2}$$

$$= \frac{r^2\sin^2 b}{\cos^2 b} + \frac{r^2\sin^2 c}{\cos^2 c} - \frac{2\,r^2\sin b\sin c\cos A}{\cos b\cos c} \qquad \text{(equation \# 3)}$$

Fourth, we consider the triangle,

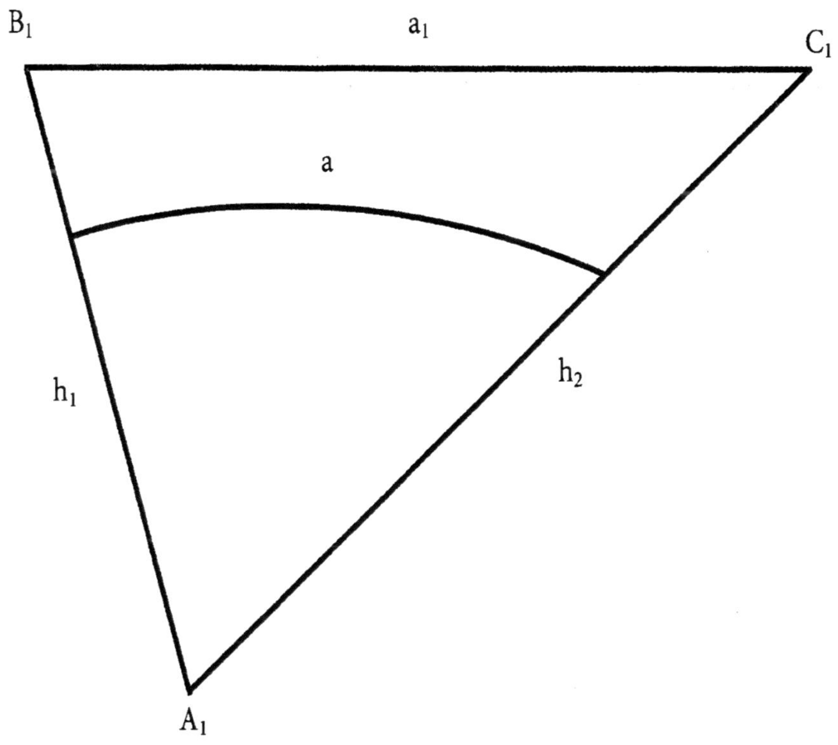

Figure # 5

Once again, using the law cosines on the plane, ${a_1}^2 = {h_2}^2 + {h_1}^2 - 2h_2h_1 \cos a$

$$= \left(\frac{r}{\cos b}\right)^2 + \left(\frac{r}{\cos c}\right)^2 - 2\left(\frac{r}{\cos b}\right)\left(\frac{r}{\cos c}\right)\cos a$$

$$= \frac{r^2}{\cos^2 b} + \frac{r^2}{\cos^2 c} - \frac{2\, r^2 \cos a}{\cos b \cos c} \qquad \text{(equation \# 4)}$$

We now set the result of equation # 3 equal to the result of equation # 4,

$$\frac{r^2 \sin^2 b}{\cos^2 b} + \frac{r^2 \sin^2 c}{\cos^2 c} - \frac{2\, r^2 \sin b \sin c \cos A}{\cos b \cos c} = \frac{r^2}{\cos^2 b} + \frac{r^2}{\cos^2 c} - \frac{2\, r^2 \cos a}{\cos b \cos c}$$

After multiplying both sides of the newly formed equation by $\cos^2 b \cos^2 c$ we obtain,

$$r^2\sin^2 b\cos^2 c + r^2\sin^2 c\cos^2 b - 2r^2\sin b\sin c\cos b\cos c\cos A$$
$$= r^2\cos^2 c + r^2\cos^2 b - 2r^2\cos b\cos c\cos a$$

Next, we divide both sides of the equation by r^2 to obtain

$$\sin^2 b\cos^2 c + \sin^2 c\cos^2 b - 2\sin b\sin c\cos b\cos c\cos A$$
$$= \cos^2 c + \cos^2 b - 2\cos b\cos c\cos a$$

Next, we isolate the term containing $\cos a$ on the left side of the equation,

$$2\cos b\cos c\cos a + \sin^2 b\cos^2 c + \sin^2 c\cos^2 b - 2\sin b\sin c\cos b\cos c\cos A$$
$$= \cos^2 c + \cos^2 b$$

$$2\cos b\cos c\cos a$$
$$= \cos^2 c + \cos^2 b - \sin^2 b\cos^2 c - \sin^2 c\cos^2 b + 2\sin b\sin c\cos b\cos c\cos A$$

$$2\cos b\cos c\cos a$$

$$= \cos^2 c - \sin^2 b\cos^2 c + \cos^2 b - \sin^2 c\cos^2 b + 2\sin b\sin c\cos b\cos c\cos A$$

$$2\cos b\cos c\cos a = \cos^2 c(1 - \sin^2 b) + \cos^2 b(1 - \sin^2 c) + 2\sin b\sin c\cos b\cos c\cos A$$

We now utilize the Pythagorean identity, $\sin^2\theta + \cos^2\theta = 1$, as follows:

$$2\cos b\cos c\cos a = \cos^2 c\cos^2 b + \cos^2 b\cos^2 c + 2\sin b\sin c\cos b\cos c\cos A$$

$$2\cos b\cos c\cos a = 2\cos^2 b\cos^2 c + 2\sin b\sin c\cos b\cos c\cos A$$

Finally, we divide both sides of the equation by $2\cos b\cos c$ to obtain

$$\cos a = \cos b\cos c + \sin b\sin c\cos A$$

which is the law of cosines for sides in spherical trigonometry.

A Closer Look at the First Example

In first example of Chapter two, the approximate distance between Malabo, Republic of Equatorial Guinea, and Pretoria, South Africa is calculated. First, the geographic coordinates of both capitals re converted into a degree decimal format where each coordinate is rounded to four decimal places:

Malabo, Republic of Equatorial Guinea is located at $3^{\circ}45' \text{ N} = \left(3 + {}^{45}/_{60}\right)^{\circ} \text{ N} = 3.7500^{\circ} \text{ N}$,

$^{\circ}47' \text{ E} = \left(8 + {}^{47}/_{60}\right)^{\circ} \text{ E} \approx 8.7833^{\circ} \text{ E}$.

Pretoria, South Africa is located at $25^{\circ}42' \text{ S} = \left(25 + {}^{42}/_{60}\right)^{\circ} \text{ S} = 25.7000^{\circ} \text{ S}$,

$8^{\circ}13' \text{ E} = \left(28 + {}^{13}/_{60}\right)^{\circ} \text{ E} \approx 28.2167^{\circ} \text{ E}$.

To create a spherical triangle, we measure the angle of the side that extends from the North Pole to Malabo. Since Malabo lies in the northern hemisphere, we subtract 3.7500° from 90°. Next, to find the measure of the side that extends from the North Pole to Pretoria, we add 25.7000° to 90° since Pretoria lies in the southern hemisphere. It follows that the two sides of the spherical triangle are $90^{\circ} - 3.7500^{\circ} = 86.2500^{\circ}$ and $90^{\circ} + 25.7000^{\circ} = 115.7000^{\circ}$, respectively. The angle between the wo sides at the North Pole is $28.2167^{\circ} - 8.7833^{\circ} = 19.4334^{\circ}$. See the illustration on page 26.

Next, we then use the spherical law of cosines to determine the cosine of side n, and then apply the inverse cosine function to calculate the measure of angle n in degrees:

$$\text{os n} = \cos(86.2500^{\circ})\cos(115.7000^{\circ}) + \sin(86.2500^{\circ})\sin(115.7000^{\circ})\cos(19.4334^{\circ})$$

$$\text{os n} \approx 0.819559619888 \Rightarrow \text{n} \approx \cos^{-1}(0.819559619888) \Rightarrow \text{n} \approx 34.959265717^{\circ}.$$

In our final step, we convert the measure of angle n from degrees to radians, and apply the arc length formula to approximate distance between the two capitals,

$$34.959265717^{\circ})\left({}^{\pi}/_{180^{\circ}}\right)(3959 \text{ miles})$$

$$\approx (0.610154290841)(3959 \text{ miles})$$

$$\approx 2{,}416 \text{ miles.}$$

Creating the illustrations

With regards to the illustrations I have drawn in this activity book, they are not spherical triangles drawn to scale. The purpose of the illustrations is to merely provide a visual aid throughout the calculations. To create the illustrations, I used a curve ruler to help depict their three dimensional nature.

To help you create your spherical triangles, I have provided seven illustrations on the following pages. Before examining the illustrations, note that E represents a longitude that is east of the prime meridian; W represents a longitude west of the prime meridian; N represents a latitude north of the equator and S represents a latitude south of the equator. The subscripts identify the city being referenced. For example, E_1 represents a longitude east of the prime meridian for city B, whereas, N_2 represents a latitude north of the equator for city C. If you examine the notes below each of the seven illustrations, you will see how the measure of the angle at the pole, P, as well as the length of side's b and c are calculated. Finally note, it is important to remember that in spherical trigonometry the sides of a spherical triangle can always be thought of in two ways. They are angles of a given measure, and they are sides of a given length.

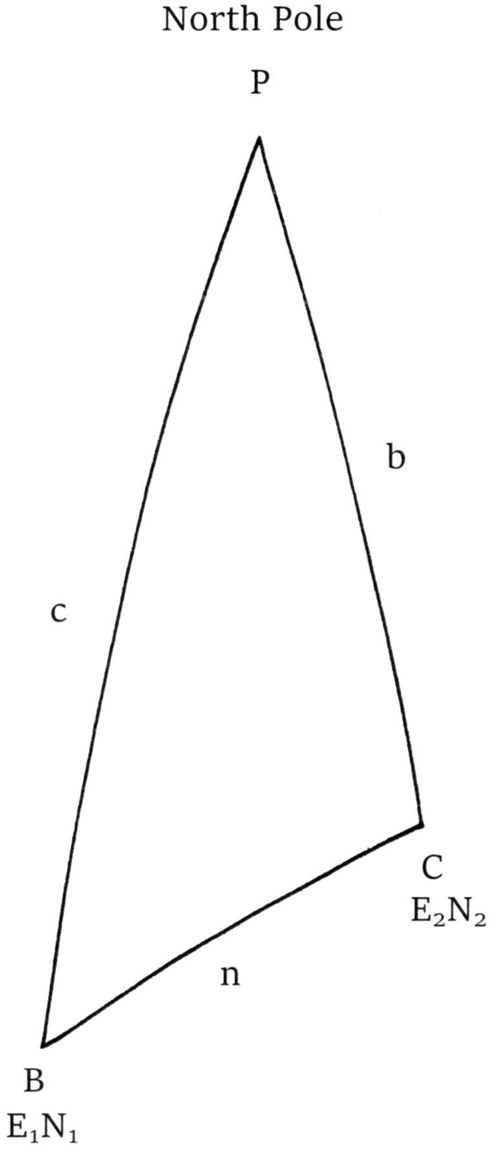

where $E_2 > E_1$

$P = E_2 - E_1$

$c = 90^o - N_1$

$b = 90^o - N_2$

Figure # 6

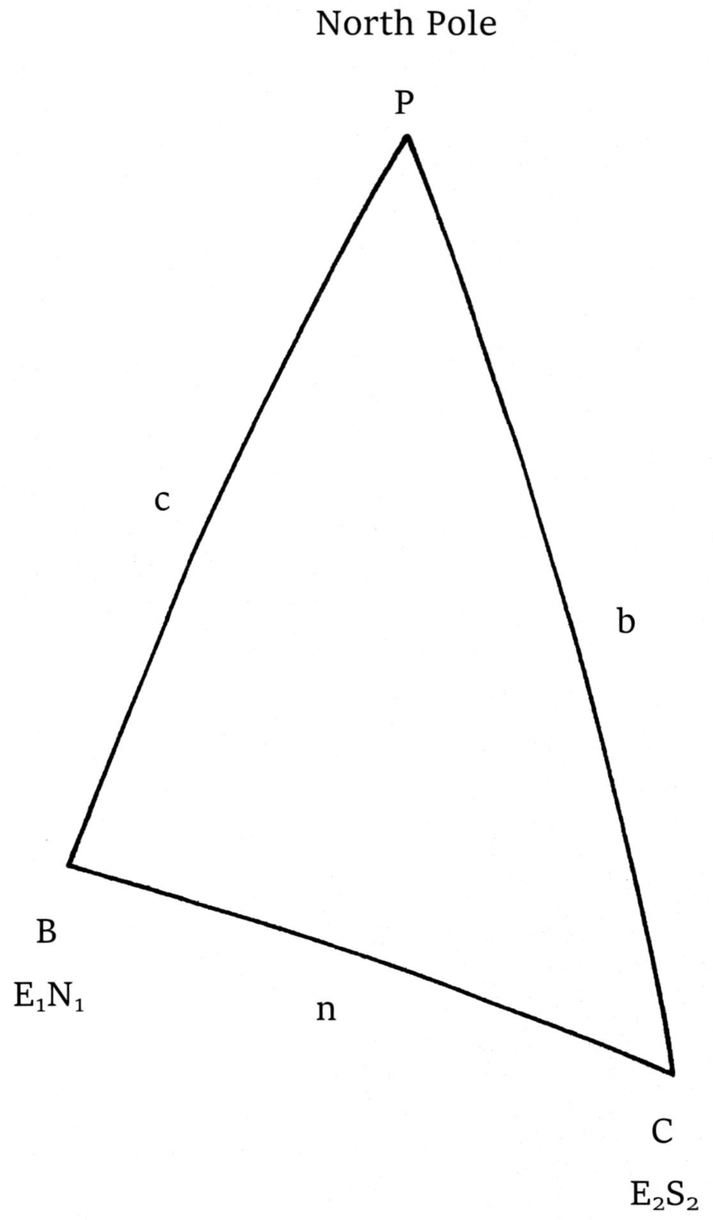

where $E_2 > E_1$

$P = E_2 - E_1$

$c = 90^o - N_1$

$b = 90^o + S_2$

Figure # 7

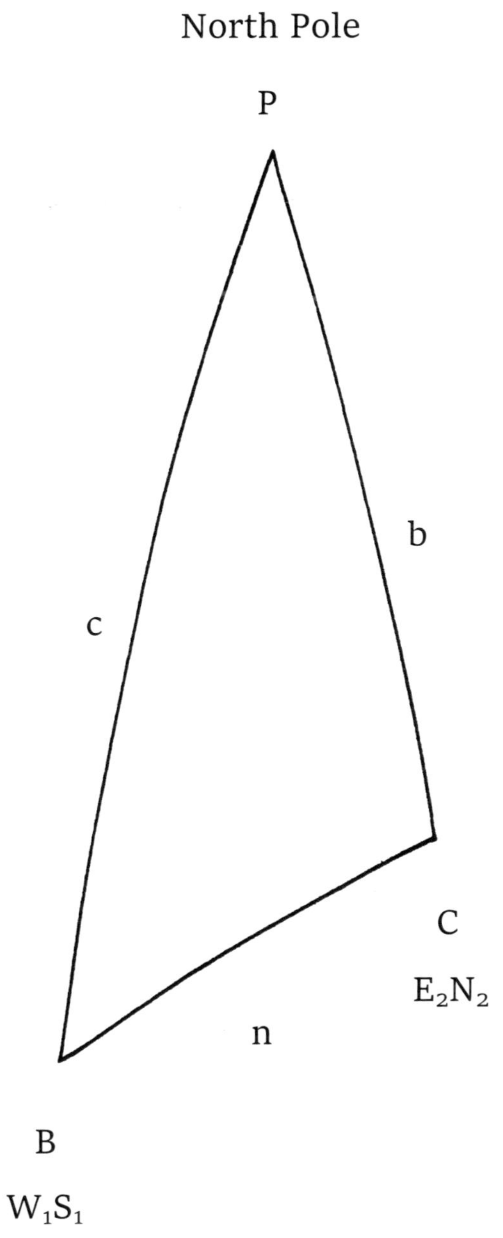

$P = W_1 + E_2$

$c = 90^{o} + S_1$

$b = 90^{o} - N_2$

Figure # 8

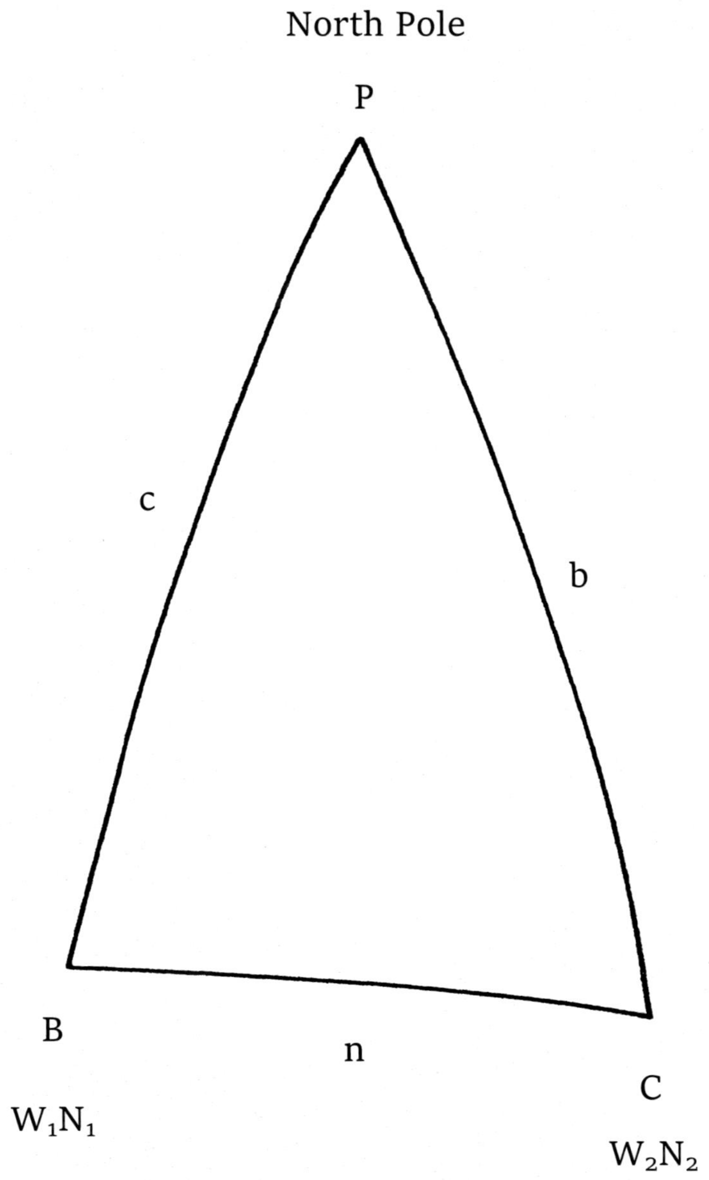

where $W_1 > W_2$

$P = W_1 - W_2$

$c = 90^o - N_1$

$b = 90^o - N_2$

Figure # 9

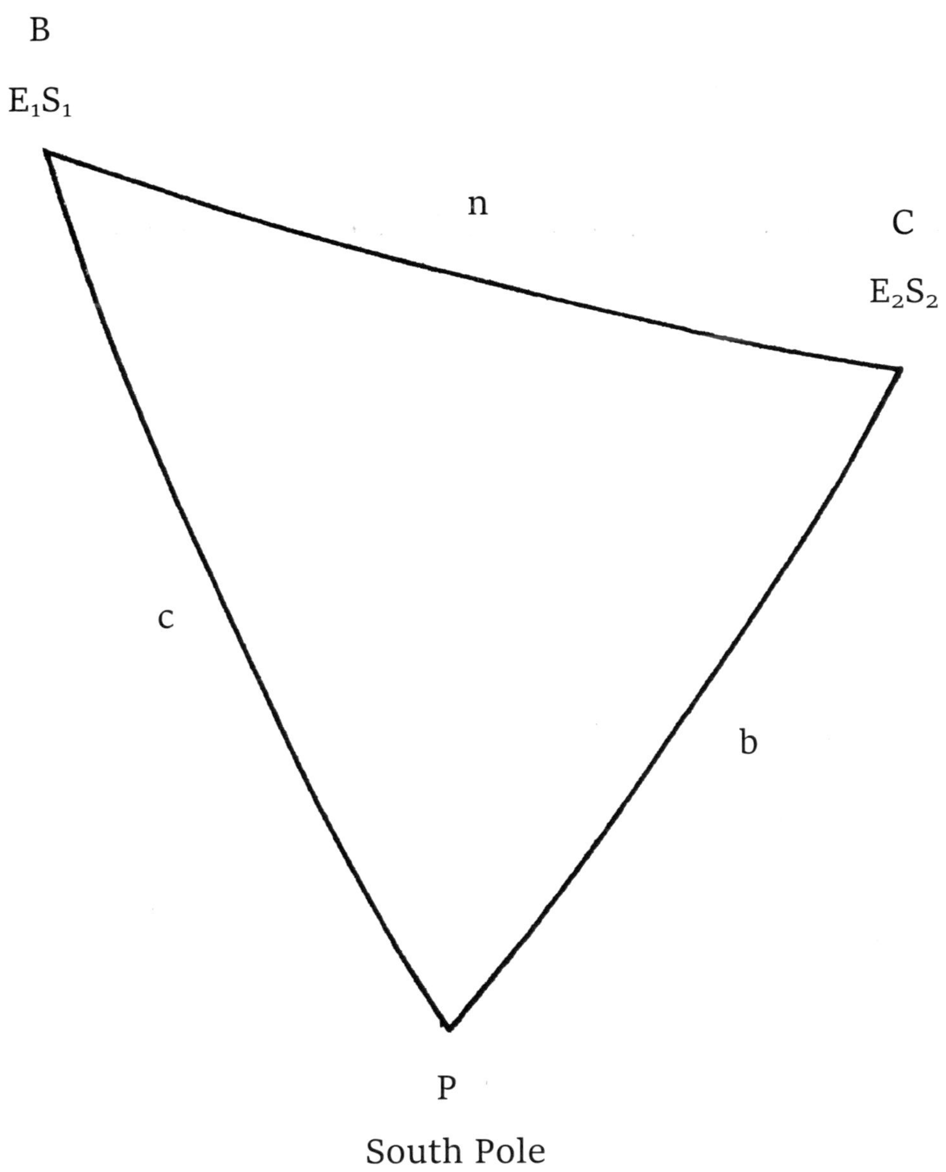

where $E_2 > E_1$

$P = E_2 - E_1$

$c = 90^o - S_1$

$b = 90^o - S_2$

Figure # 10

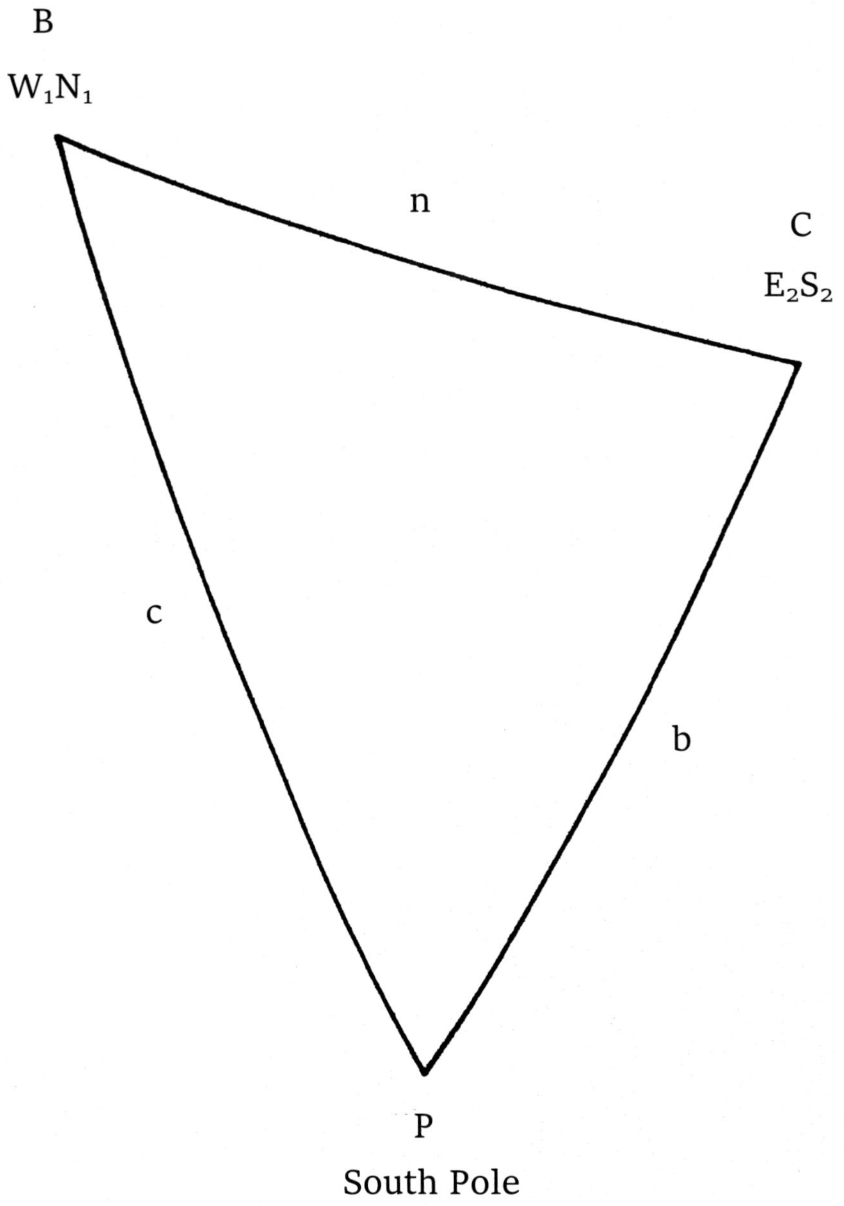

$P = W_1 + E_2$

$c = 90^o + N_1$

$b = 90^o - S_2$

Figure # 11

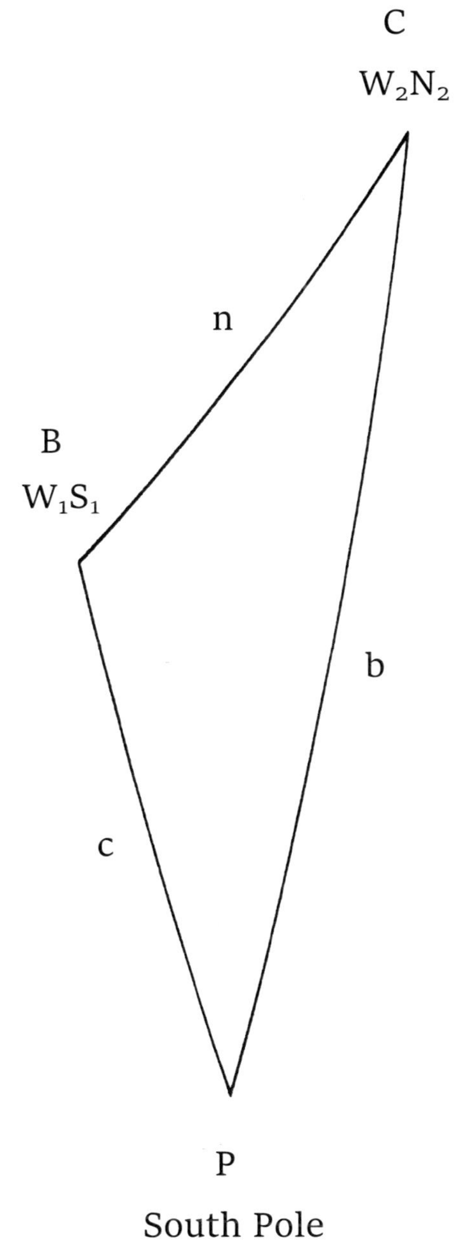

where $W_1 > W_2$

$P = W_1 - W_2$

$c = 90^0 - S_1$

$b = 90^0 + N_2$

Figure # 12

Chapter 2
Examples & Problems

Example 1

Malabo, Republic of Equatorial Guinea is located at

$3^o 45'\ \mathrm{N} = \left(3 + {}^{45}/_{60}\right)^o\ \mathrm{N} = 3.7500^o\ \mathrm{N}\,,\ 8^o 47'\ \mathrm{E} = \left(8 + {}^{47}/_{60}\right)^o\ \mathrm{E} \approx 8.7833^o\ \mathrm{E}\,.$

Pretoria, South Africa is located at

$25^o 42'\ \mathrm{S} = \left(25 + {}^{42}/_{60}\right)^o\ \mathrm{S} = 25.7000^o\ \mathrm{S}\ ,\ 28^o 13'\ \mathrm{E} = \left(28 + {}^{13}/_{60}\right)^o\ \mathrm{E} \approx 28.2167^o\ \mathrm{E}\,.$

Two sides of the spherical triangle are $90^o - 3.7500^o = 86.2500^o$ and $90^o + 25.7000^o = 115.7000^o$, respectively. The angle at the North Pole is $28.2167^o - 8.7833^o = 19.4334^o$.

See the illustration on the next page.

It follows that

$\cos \mathrm{n} = \cos(86.2500^o)\cos(115.7000^o) + \sin(86.2500^o)\ \sin(115.7000^o)\cos(19.4334^o)$

$\cos \mathrm{n} \approx 0.819559619888 \Rightarrow \mathrm{n} \approx \cos^{-1}(0.819559619888) \Rightarrow \mathrm{n} \approx 34.959265717^o.$

Therefore, the approximate distance between the two capital cities is

$(34.959265717^o)\left({}^{\pi}/_{180^o}\right)(3959 \text{ miles})$

$\approx (0.610154290841)(3959 \text{ miles})$

$\approx\ 2{,}416 \text{ miles}\,.$

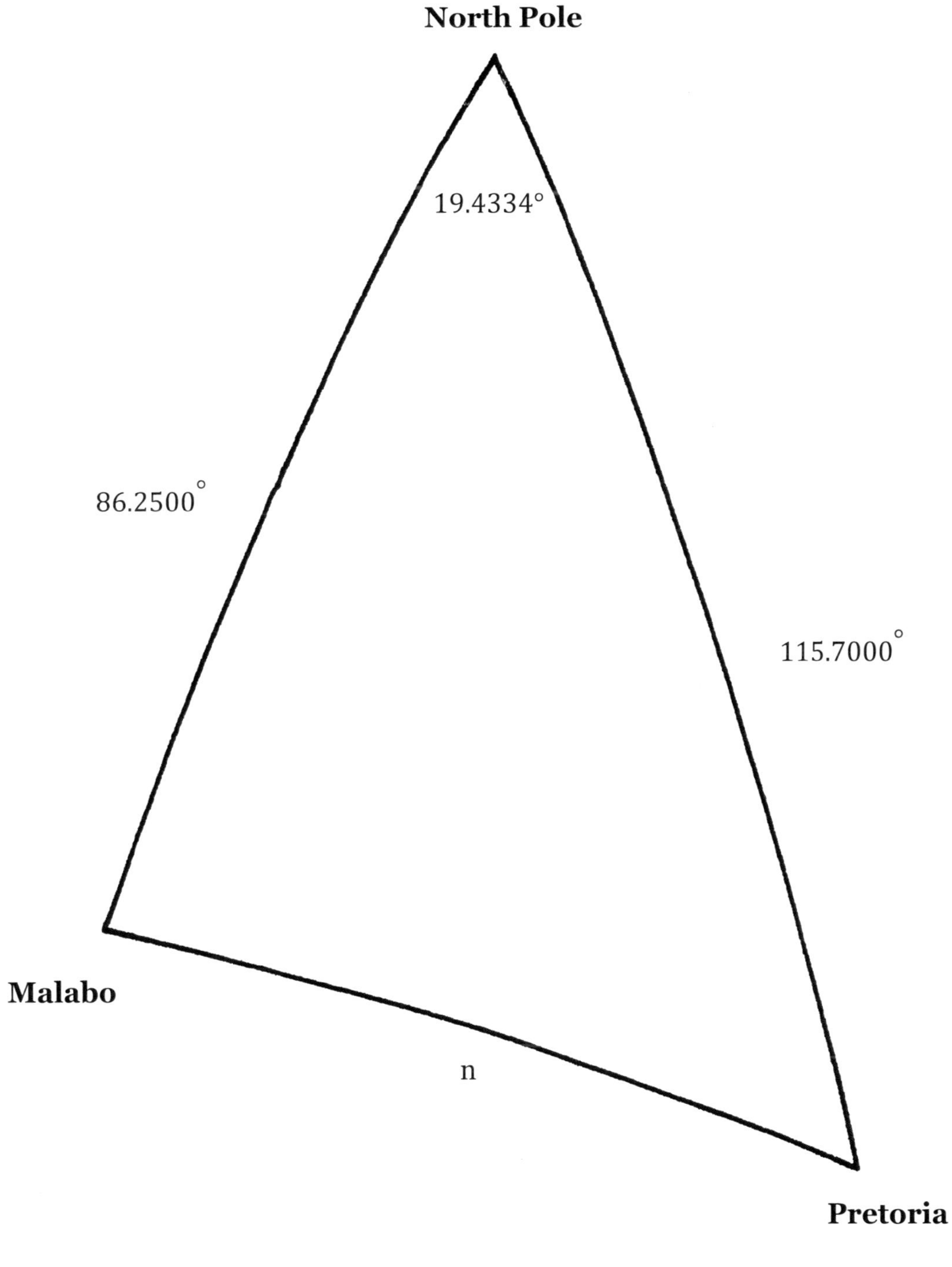
North Pole
19.4334°
86.2500°
115.7000°
Malabo
n
Pretoria

Example 2

Bissau, Guinea-Bissau is located at $11^{o}51'$ N $= \left(11 + {}^{51}/_{60}\right)^{o}$ N $= 11.8500^{o}$ N ,

$15^{o}35'$ W $= \left(15 + {}^{35}/_{60}\right)^{o}$ W $\approx 15.5833^{o}$ W .

Dar es Salaam, Tanzania is located at $6^{o}48'$ S $= \left(6 + {}^{48}/_{60}\right)^{o}$ S $\approx 6.8000^{o}$ S ,

$39^{o}17'$ E $= \left(39 + {}^{17}/_{60}\right)^{o}$ E $\approx 39.2833^{o}$ E .

Two sides of the spherical triangle are $90^{o} + 11.8500^{o} = 101.8500^{o}$ and $90^{o} - 6.8000^{o} = 83.2000^{o}$, respectively. The angle at the South Pole is $15.5833^{o} + 39.2833^{o} = 54.8666^{o}$.

See the illustration on the next page.

It follows that

$\cos n = \cos(101.8500^{o})\cos(83.2000^{o}) + \sin(101.8500^{o})\ \sin(83.2000^{o})\cos(54.8666^{o})$

$\cos n \approx 0.534941505514 \quad \Rightarrow n \approx \cos^{-1}(0.534941505514\ ) \Rightarrow n \approx 57.6600554513^{o}$.

Therefore, the approximate distance between the two capital cities is

$(57.6600554513^{o})({}^{\pi}/_{180^{o}})(3959 \text{ miles})$

$\approx (1.00635781451)(3959 \text{ miles})$.

$\approx$ 3,984 miles.

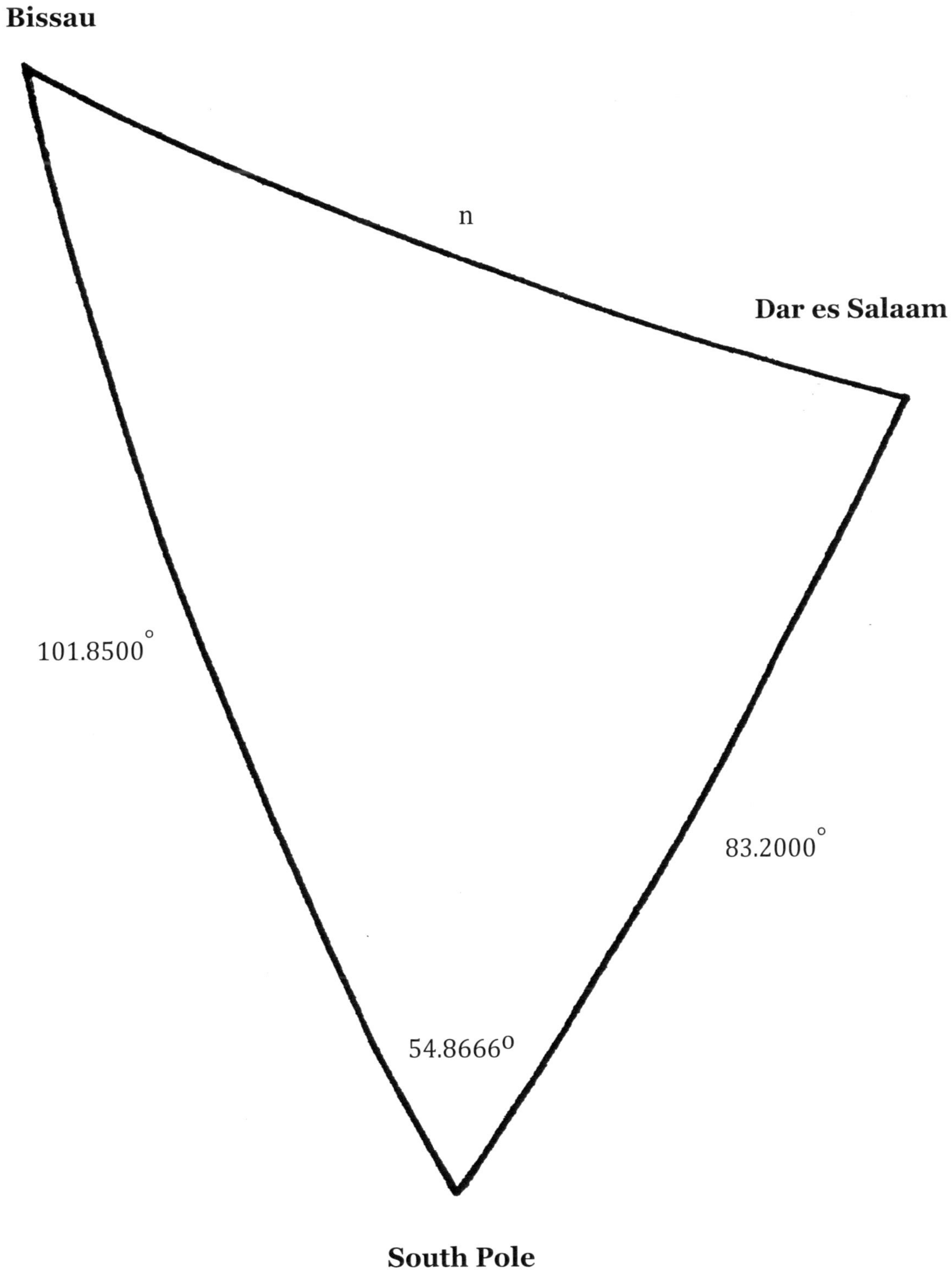
Bissau
n
Dar es Salaam
101.8500°
83.2000°
54.8666°
South Pole

Example 3

Mbabane, Swaziland is located at $26^{o}19'\ \mathrm{S} = \left(26 + {}^{19}/_{60}\right)^{o}\mathrm{S} \approx 26.3167^{o}\mathrm{S}$,

$31^{o}8'\ \mathrm{E} = \left(31 + {}^{8}/_{60}\right)^{o}\mathrm{E} \approx 31.1333^{o}\mathrm{E}$.

Antananarivo, Madagascar is located at $18^{o}55'\ \mathrm{S} = \left(18 + {}^{55}/_{60}\right)^{o}\mathrm{S} \approx 18.9167^{o}\mathrm{S}$,

$47^{o}31'\ \mathrm{E} = \left(47 + {}^{31}/_{60}\right)^{o}\mathrm{E} \approx 47.5167^{o}\mathrm{E}$.

Two sides of the spherical triangle are $90^{o} - 26.3167^{o} = 63.6833^{o}$ and $90^{o} - 18.9167^{o} = 71.0833^{o}$, respectively. The angle at the South Pole is $47.5167^{o} - 31.1333^{o} = 16.3834^{o}$.

See the illustration on the next page.

It follows that

$\cos \mathrm{n} = \cos(63.6833^{o})\cos(71.0833^{o}) + \sin(63.6833^{o})\ \sin(71.0833^{o})\cos(16.3834^{o})$

$\cos \mathrm{n} \approx\ 0.957240946572 \Rightarrow\ \mathrm{n} \approx \cos^{-1}(0.957240946572)\ \Rightarrow\ \mathrm{n} \approx 16.8155644284^{o}$.

Therefore, the approximate distance between the two capital cities is
$(16.8155644284^{o})\left({}^{\pi}/_{180^{o}}\right)(3959\ \text{miles})$

$\approx (0.293486964856)(3959\ \text{miles})$

$\approx\ 1{,}162$ miles.

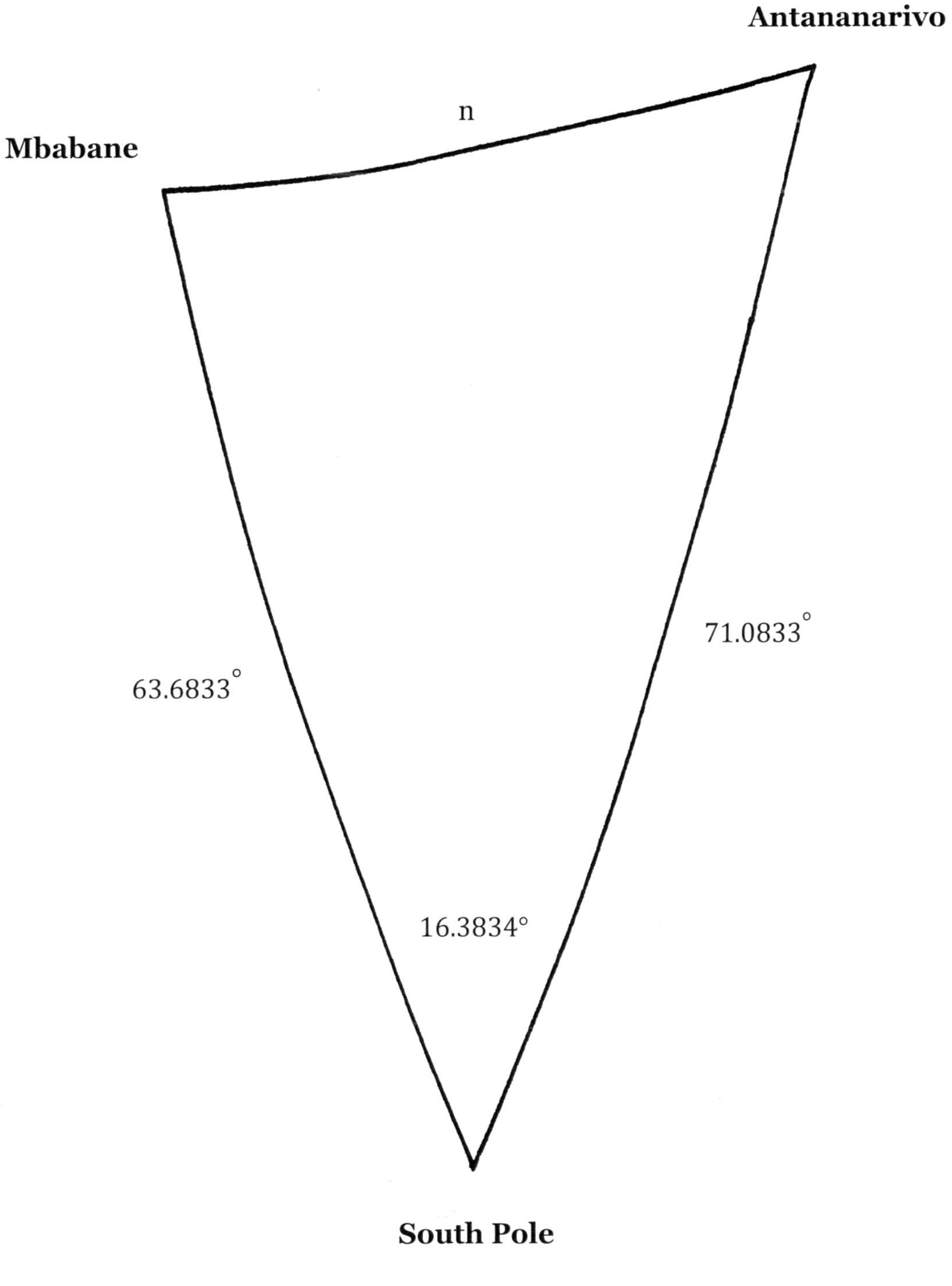
Antananarivo
n
Mbabane
71.0833°
63.6833°
16.3834°
South Pole

Chapter 2 Problems:

Approximate the distance between the following capitals in Africa.

1. Algiers, Algeria $(36^\circ 45' \text{ N } 3^\circ 3' \text{ E})$
   to Addis Ababa, Ethiopia $(9^\circ 2' \text{ N } 38^\circ 42' \text{ E})$

2. Luanda, Angola (8°50′ S 13°13′ E)

to Asmara, Eritrea (15°20′ N 38°56′ E)

3. Porto-Novo, Benin (6°29′ N 2°37′ E)

to Libreville, Gabon (0°23′ N 9°27′ E)

4. Gaborone, Botswana (24°38′ S 25°54′ E)

to Harare, Zimbabwe (17°49′ S 31°2′ E)

5. Ouagadougou, Burkina Faso (12°22′ N 1°31′ W) to Kampala, Uganda (0°19′ N 32°33′ E)

6. Yaoundé, Cameroon ($3^{\circ}52'$ N $11^{\circ}31'$ E)
to Lusaka, Zambia ($15^{\circ}25'$ S $28^{\circ}17'$ E)

7. Praia, Cabo Verde (14°55′ N 23°31′ W)
to N'Djamena, Chad (12°6′ N 15°2′ E)

8. Yamoussoukro, Cote d'Ivoire (6°49′ N 5°16′ W)
to Tripoli, Libya (32°53′ N 13°10′ E)

9. Kinshasa, Democratic Republic of the Congo (4°19′ S 15°18′ E) to Nairobi, Kenya (1°17′S 36°49′E)

10. Banjul, The Gambia (13°27′ N 16°34′ W)

to São Tomé, São Tomé and Principe (0°20′ N 6°44′ E)

11. Accra, Ghana (5°33′ N 0°13′ W)

to Moroni, Comoros (11°42′ S 43°14′ E)

12. Conakry, Guinea (9°30′ N 13°42′ W)

to Abuja, Nigeria (9°5′ N 7°32′ E)

13. Maseru, Lesotho (29°19′ S 27°29′ E)

to Lilongwe, Malawi (13°58′ S 33°47′ E)

14. Monrovia, Liberia ($6^{\circ}18'$ N $10^{\circ}48'$ W)

to Khartoum, Sudan ($15^{\circ}36'$ N $32^{\circ}32'$ E)

15. Bamako, Mali (12°39′ N 8°0′ W)

to Bangui, Central African Republic (4°22′ N 18°35′ E)

16. Nouakchott, Mauritania (18°4′ N 15°58′ W)

to Victoria, Seychelles (4°37′ S 55°27′ E)

17. Rabat, Morocco (34°1′ N 6°49′ W)

to Windhoek, Namibia (22°34′ S 17°5′ E)

18. Niamey, Niger (13°31′ N 2°7′ E)

to Kigali, Rwanda (1°57′ S 30°3′ E)

19. Brazzaville, Republic of the Congo (4°15′ S 15°17′ E) to Djibouti, Djibouti (11°35′ N 43°9′ E)

20. Dakar, Senegal (14°44′ N 17°38′ W)

to Cairo, Egypt (30°3′ N 31°15′ E)

21. Freetown, Sierra Leone (8°29′ N 13°14′ W)

to Bujumbura, Burundi (3°22′ S 29°21′ E)

22. Juba, South Sudan (4°51′ N 31°37′ E)

to Mogadishu, Somalia (2°4′ N 45°20′ E)

23. Lomé, Togo (6°7′ N 1°13′ E)

to Maputo, Mozambique (25°57′ S 32°35′ E)

24. Tunis, Tunisia (36°48′ N 10°11′ E)

to Port Louis, Mauritius (20°9′ S 57°29′ E)

Chapter 3

Solutions

Chapter 2

Problem 1

Solution

Algiers, Algeria is located at $36^o45' \text{ N} = \left(36 + {}^{45}/_{60}\right)^o \text{ N} = 36.7500^o \text{ N}$, $3^o3' \text{ E} = \left(3 + {}^{3}/_{60}\right)^o \text{ E} = 3.0500^o \text{ E}$.

Addis Ababa, Ethiopia is located at $9^o2' \text{ N} = \left(9 + {}^{2}/_{60}\right)^o \text{ N} \approx 9.0333^o \text{ N}$, $38^o42' \text{ E} = \left(38 + {}^{42}/_{60}\right)^o \text{ E} = 38.7000^o \text{ E}$.

Two sides of the spherical triangle are $90^o - 36.7500^o = 53.2500^o$ and $90^o - 9.0333^o = 80.9667^o$, respectively. The angle at the North Pole is $38.7000^o - 3.0500^o = 35.6500^o$.

See the illustration on the next page.

It follows that

$\cos \text{n} = \cos(53.2500^o)\cos(80.9667^o) + \sin(53.2500^o)\ \sin(80.9667^o)\cos(35.6500^o)$

$\cos \text{n} \approx 0.736959498133 \Rightarrow \text{n} \approx \cos^{-1}(0.736959498133) \Rightarrow \text{n} \approx 42.5269485004^o$.

Therefore, the approximate distance between the two capital cities is

$(42.5269485004^o)\left({}^{\pi}/_{180^o}\right)(3959 \text{ miles})$

$\approx (0.742235272158)(3959 \text{ miles})$

$\approx$ 2,939 miles.

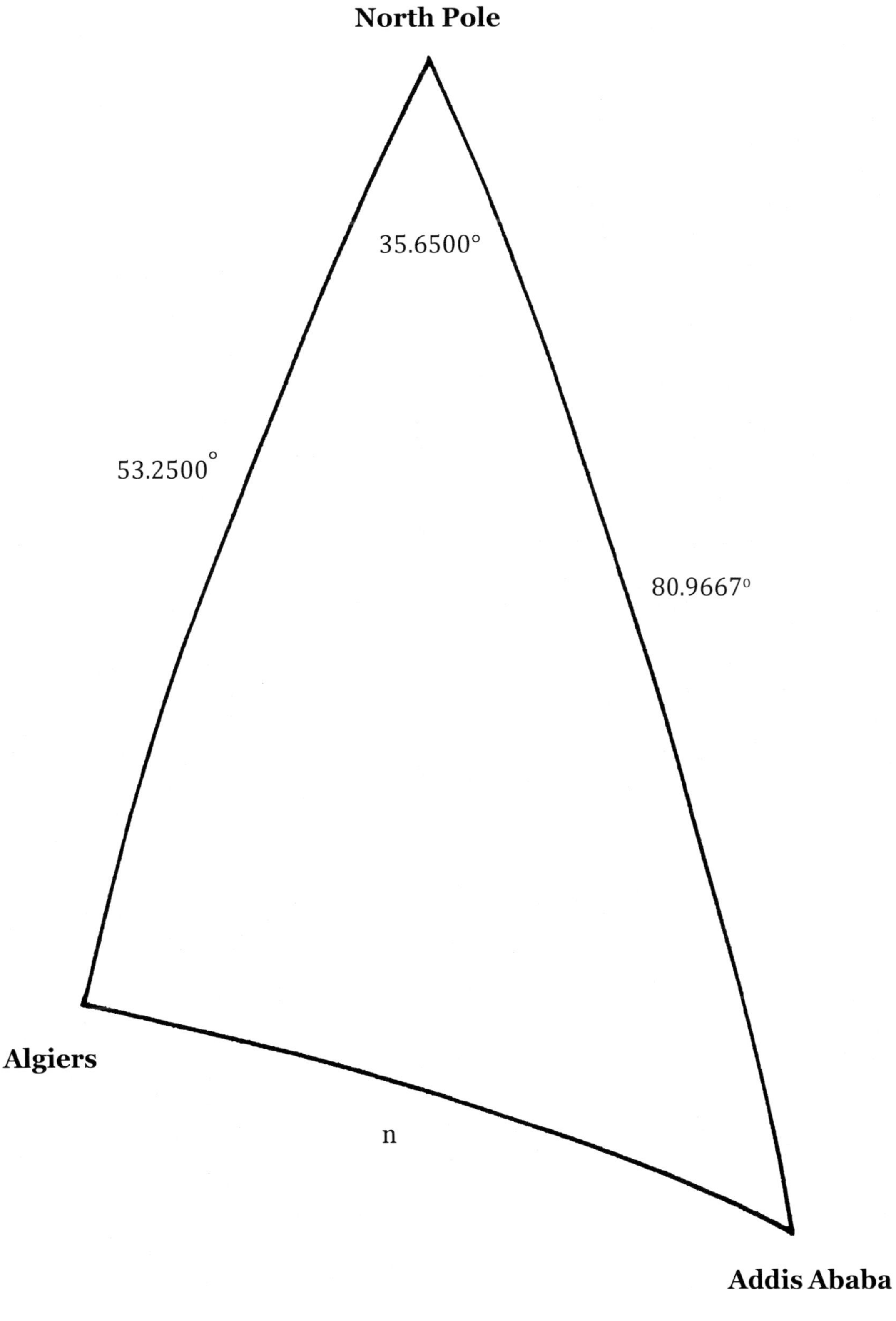
North Pole
35.6500°
53.2500°
80.9667°
Algiers
n
Addis Ababa

Chapter 2
Problem 2
Solution

Luanda, Angola is located at $8^o 50' \text{ S} = \left(8 + {}^{50}/_{60}\right)^o \text{ S} \approx 8.8333^o \text{ S}$,
$13^o 13' \text{ E} = \left(13 + {}^{13}/_{60}\right)^o \text{ E} \approx 13.2167^o \text{ E}$.

Asmara, Eritrea is located at $15^o 20' \text{ N} = \left(15 + {}^{20}/_{60}\right)^o \text{ N} \approx 15.3333^o \text{ N}$, $38^o 56'$
$\text{E} = \left(38 + {}^{56}/_{60}\right)^o \text{ E} \approx 38.9333^o \text{ E}$.

Two sides of the spherical triangle are $90^o + 8.8333^o = 98.8333^o$ and
$90^o - 15.3333^o = 74.6667^o$, respectively. The angle at the North Pole is
$38.9333^o - 13.2167^o = 25.7166^o$.

See the illustration on the next page.

It follows that

$$\cos n = \cos(98.8333^o)\cos(74.6667^o) + \sin(98.8333^o)\ \sin(74.6667^o)\cos(25.7166^o)$$

$$\cos n \approx 0.817968978619 \Rightarrow n \approx \cos^{-1}(0.817968978619) \Rightarrow n \approx 35.1180054986^o.$$

Therefore, the approximate distance between the two capital cities is
$(35.1180054986^o)\left({}^{\pi}/_{180^o}\right)(3959 \text{ miles})$

$\approx (0.612924822685)(3959 \text{ miles})$

$\approx$ 2,427 miles.

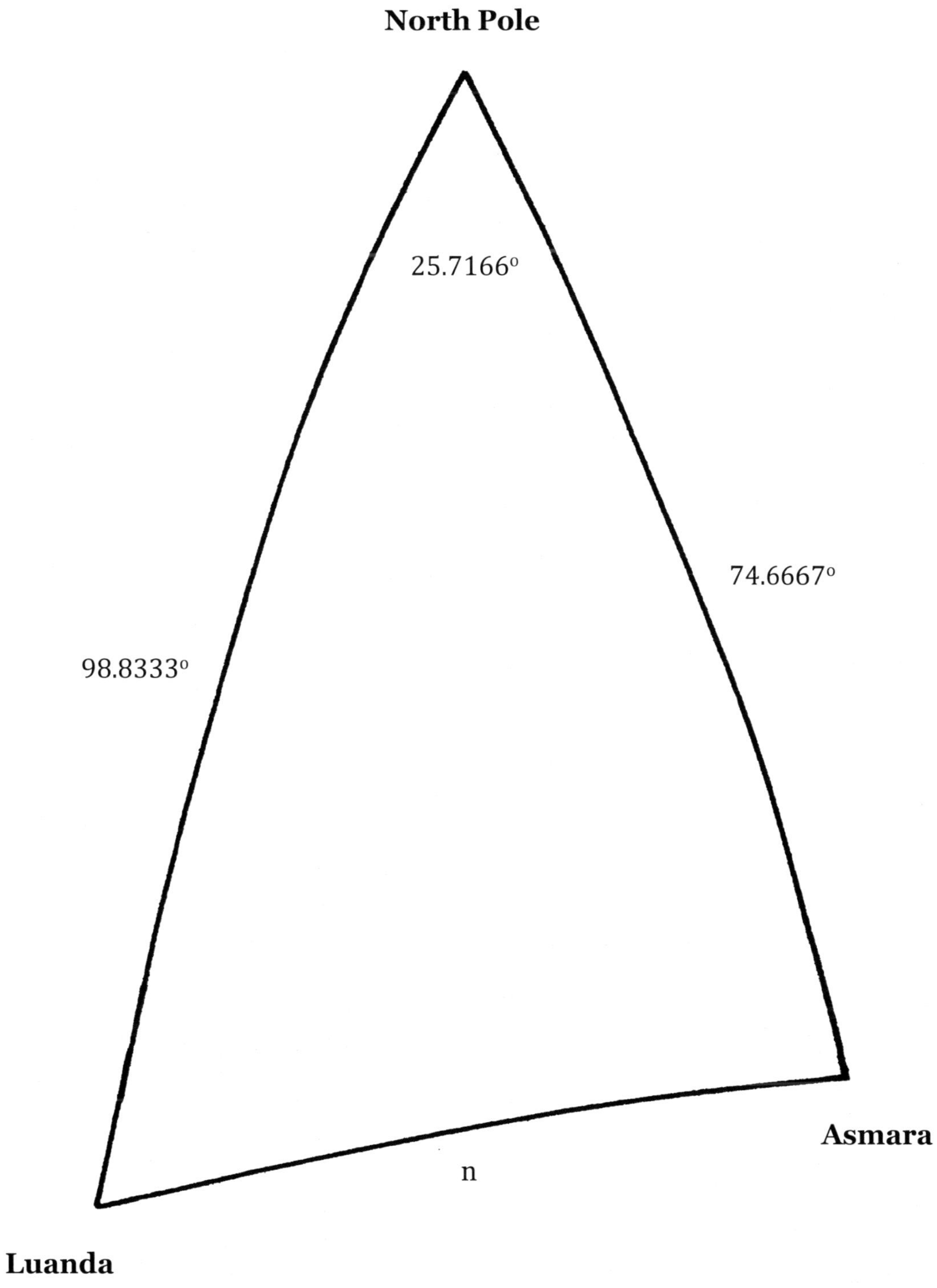
North Pole
25.7166°
74.6667°
98.8333°
Asmara
n
Luanda

Chapter 2
Problem 3
Solution

Porto-Novo, Benin is located at $6^o29'$ N $= \left(6 + {}^{29}/_{60}\right)^o$ N $\approx 6.4833^o$ N ,

$2^o37'$ E $= \left(2 + {}^{37}/_{60}\right)^o$ E $\approx 2.6167^o$ E .

Libreville, Gabon is located at $0^o23'$ N $= \left({}^{23}/_{60}\right)^o$ N $\approx 0.3833^o$ N ,

$9^o27'$ E $= \left(9 + {}^{27}/_{60}\right)^o$ E $= 9.4500^o$ E .

Two sides of the spherical triangle are $90^o - 6.4833^o = 83.5167^o$ and $90^o - 0.3833^o = 89.6167^o$, respectively. The angle at the North Pole is $9.4500^o - 2.6167^o = 6.8333^o$.

See the illustration on the next page.

It follows that

$\cos \mathrm{n} = \cos(83.5167^o)\cos(89.6167^o) + \sin(83.5167^o)\ \sin(89.6167^o)\cos(6.8333^o)$

$\cos \mathrm{n} \approx 0.987280054807 \ \Rightarrow \ \mathrm{n} \approx \cos^{-1}(0.987280054807) \ \Rightarrow \mathrm{n} \approx 9.14832695812^o.$

Therefore, the approximate distance between the two capital cities is

$(9.14832695812^o)\left({}^{\pi}/_{180^o}\right)(3959 \text{ miles})$

$\approx (0.159668426468)(3959 \text{ miles})$

$\approx\ 632\ $ miles.

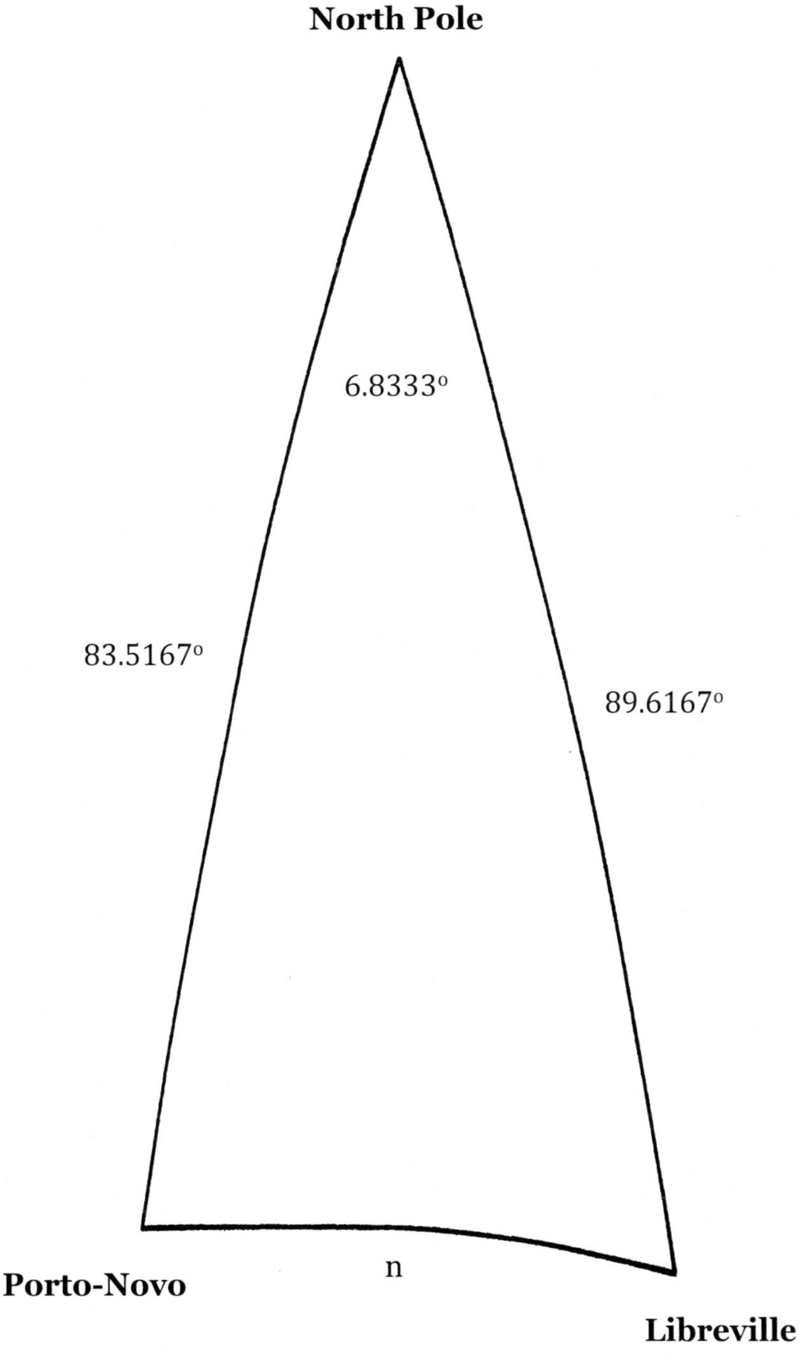
North Pole
6.8333°
83.5167°
89.6167°
n
Porto-Novo
Libreville

Chapter 2
Problem 4
Solution

Gaboronе, Botswana is located at $24^o38' \text{ S} = \left(24 + {}^{38}/_{60}\right)^o \text{ S} \approx 24.6333^o \text{ S}$, $25^o54' \text{ E} = \left(25 + {}^{54}/_{60}\right)^o \text{ E} = 25.9000^o \text{ E}$.

Harare, Zimbabwe is located at $17^o49' \text{ S} = \left(17 + {}^{49}/_{60}\right)^o \text{ S} \approx 17.8167^o \text{ S}$, $31^o2' \text{ E} = \left(31 + {}^{2}/_{60}\right)^o \text{ E} \approx 31.0333^o \text{ E}$.

Two sides of the spherical triangle are $90^o - 24.6333^o = 65.3667^o$ and $90^o - 17.8167^o \approx 72.1833^o$, respectively. The angle at the South Pole is $31.0333^o - 25.9000^o = 5.1333^o$.

See the illustration on the next page.

It follows that

$\cos \text{n} = \cos(65.3667^o)\cos(72.1833^o) + \sin(65.3667^o)\ \sin(72.1833^o)\cos(5.1333^o)$

$\cos \text{n} \approx 0.989460247964 \ \Rightarrow \ \text{n} \approx \cos^{-1}(0.989460247964) \ \Rightarrow \ \text{n} \approx 8.32597327559^o$.

Therefore, the approximate distance between the two capital cities is $(8.32597327559^o)\left({}^{\pi}/_{180^o}\right)(3959 \text{ miles})$

$\approx (0.145315647092)(3959 \text{ miles})$

$\approx\ 575\ \text{ miles}$.

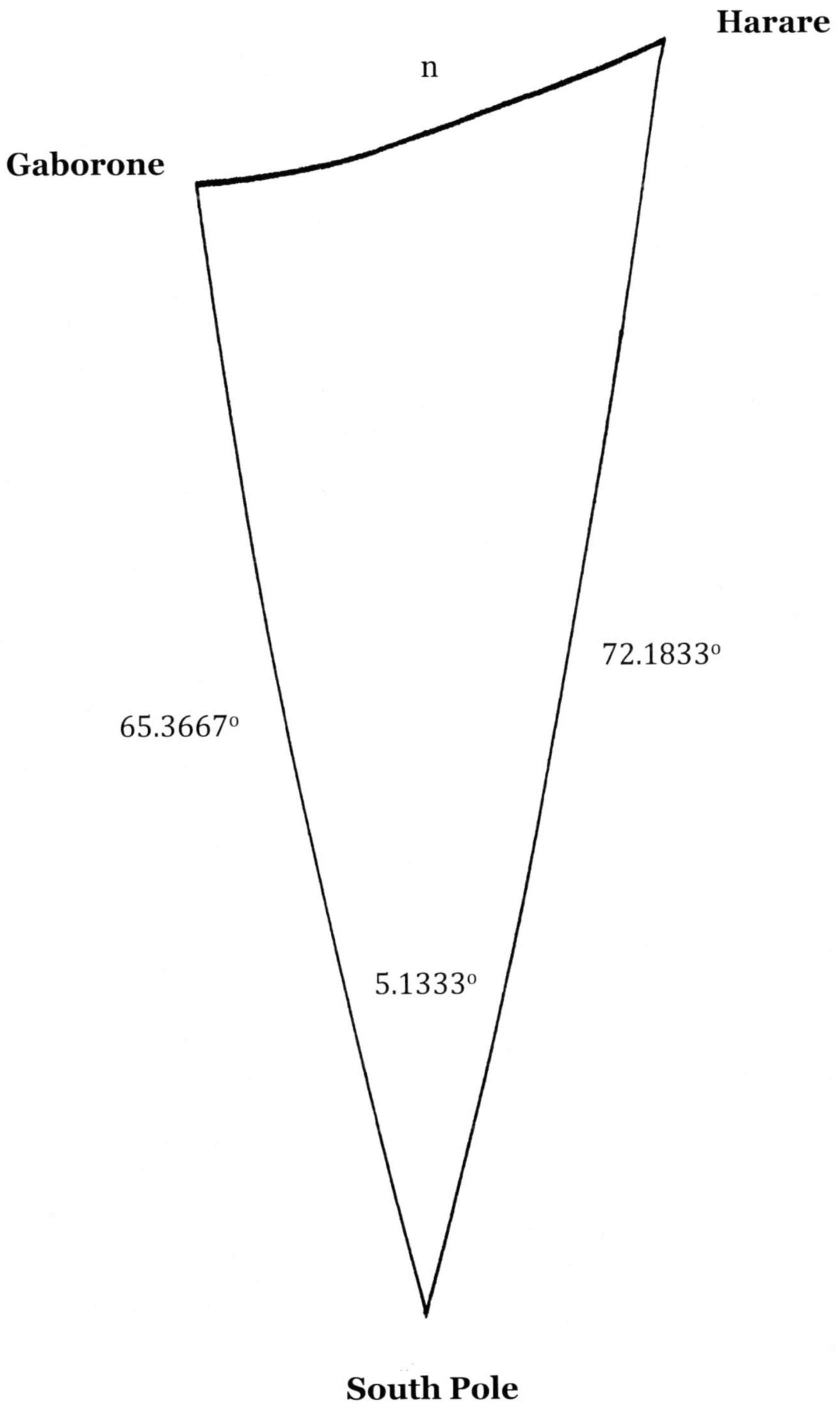
Harare
n
Gaborone
72.1833°
65.3667°
5.1333°
South Pole

Chapter 2
Problem 5
Solution

Ouagadougou, Burkina Faso is located at

$12^o22'\ \mathrm{N} = \left(12 + {}^{22}/_{60}\right)^o\ \mathrm{N} \approx 12.3667^o\ \mathrm{N}$, $1^o31'\ \mathrm{W} = \left(1 + {}^{31}/_{60}\right)^o\ \mathrm{W} \approx 1.5167^o\ \mathrm{W}$.

Kampala, Uganda is located at

$0^o19'\ \mathrm{N} = \left({}^{19}/_{60}\right)^o\ \mathrm{N} \approx 0.3167^o\ \mathrm{N}$,$32^o33'\ \mathrm{E} = \left(32 + {}^{33}/_{60}\right)^o\ \mathrm{E} \approx 32.5500^o\ \mathrm{E}$.

Two sides of the spherical triangle are $90^o - 12.3667^o = 77.6333^o$ and $90^o - 0.3167^o = 89.6833^o$, respectively. The angle at the North Pole is $1.5167^o + 32.5500^o = 34.0667^o$.

See the illustration on the next page.

It follows that

$\cos \mathrm{n} = \cos(77.6333^o)\cos(89.6833^o) + \sin(77.6333^o)\ \sin(89.6833^o)\cos(34.0667^o)$

$\cos \mathrm{n} \approx 0.810336360238 \ \Rightarrow\ \mathrm{n} \approx \cos^{-1}(0.810336360238)\ \Rightarrow\ \mathrm{n} \approx 35.8711922515^o$.

Therefore, the approximate distance between the two capital cities is
$(35.8711922515^o)\left({}^{\pi}/_{180^o}\right)(3959\ \text{miles})$

$\approx (0.626070411405)(3959\ \text{miles})$

$\approx\ 2{,}479\ \text{miles}$.

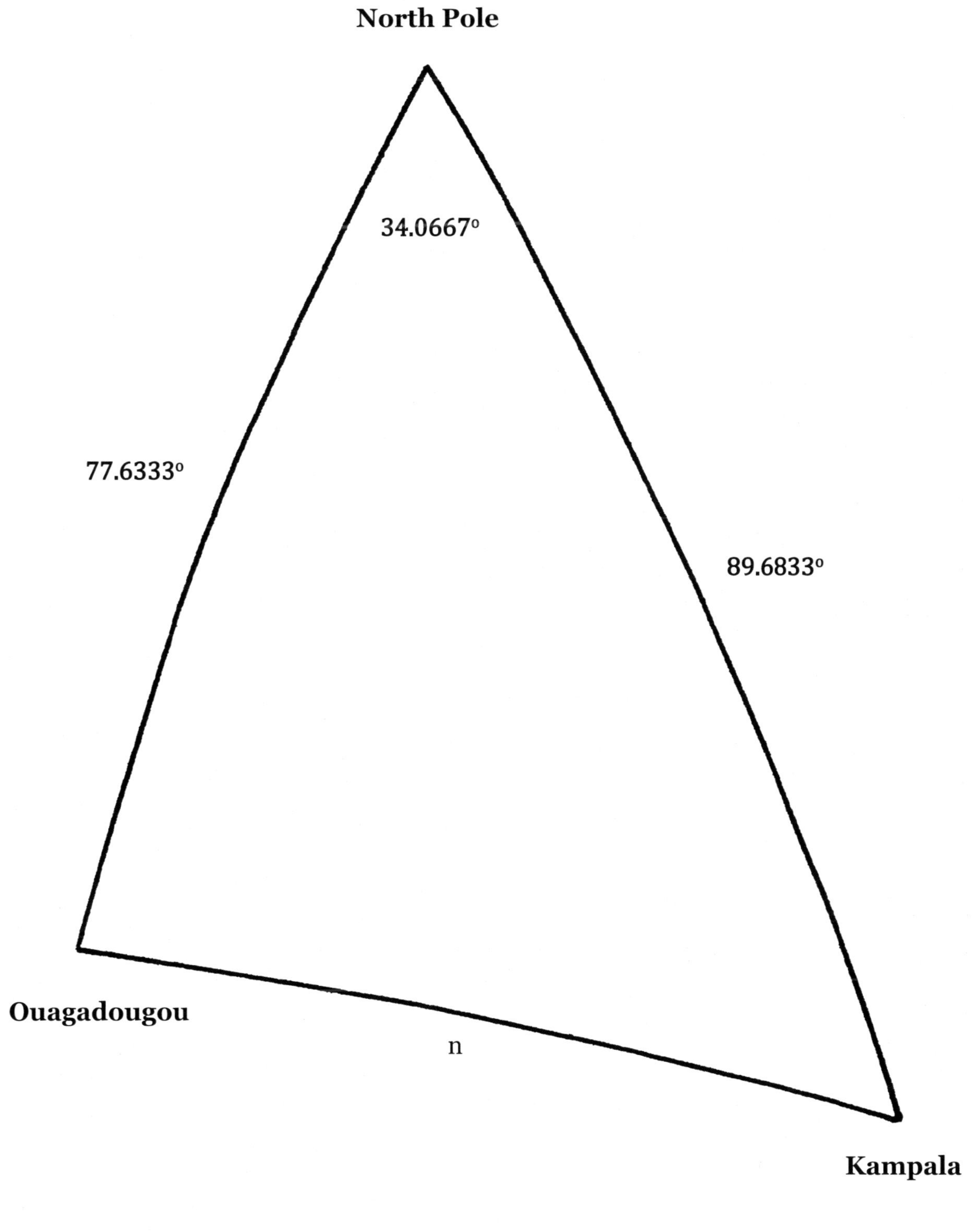
North Pole
34.0667°
77.6333°
89.6833°
Ouagadougou
n
Kampala

Chapter 2
Problem 6
Solution

Yaoundé, Cameroon is located at $3^o52'\text{ N} = \left(3 + {}^{52}/_{60}\right)^o\text{ N} \approx 3.8667^o\text{ N}$,

$11^o31'\text{ E} = \left(11 + {}^{31}/_{60}\right)^o\text{ E} \approx 11.5167^o\text{ E}$.

Lusaka, Zambia is located at $15^o25'\text{ S} = \left(15 + {}^{25}/_{60}\right)^o\text{ S} \approx 15.4167^o\text{ S}$,

$28^o17'\text{ E} = \left(28 + {}^{17}/_{60}\right)^o\text{ E} \approx 28.2833^o\text{ E}$.

Two sides of the spherical triangle are $90^o + 3.8667^o = 93.8667^o$ and $90^o - 15.4167^o = 74.5833^o$, respectively. The angle at the South Pole is $28.2833^o - 11.5167^o \approx 16.7666^o$.

See the illustration on the next page.

It follows that

$\cos n = \cos(93.8667^o)\cos(74.5833^o) + \sin(93.8667^o)\ \sin(74.5833^o)\cos(16.7666^o)$

$\cos n \approx 0.903007458883 \Rightarrow n \approx \cos^{-1}(0.903007458883) \Rightarrow n \approx 25.4437558295^o$.

Therefore, the approximate distance between the two capital cities is

$(25.4437558295^o)\left({}^{\pi}/_{180^o}\right)(3959\text{ miles})$

$\approx (0.444077313298)(3959\text{ miles})$

$\approx 1{,}758\text{ miles}$.

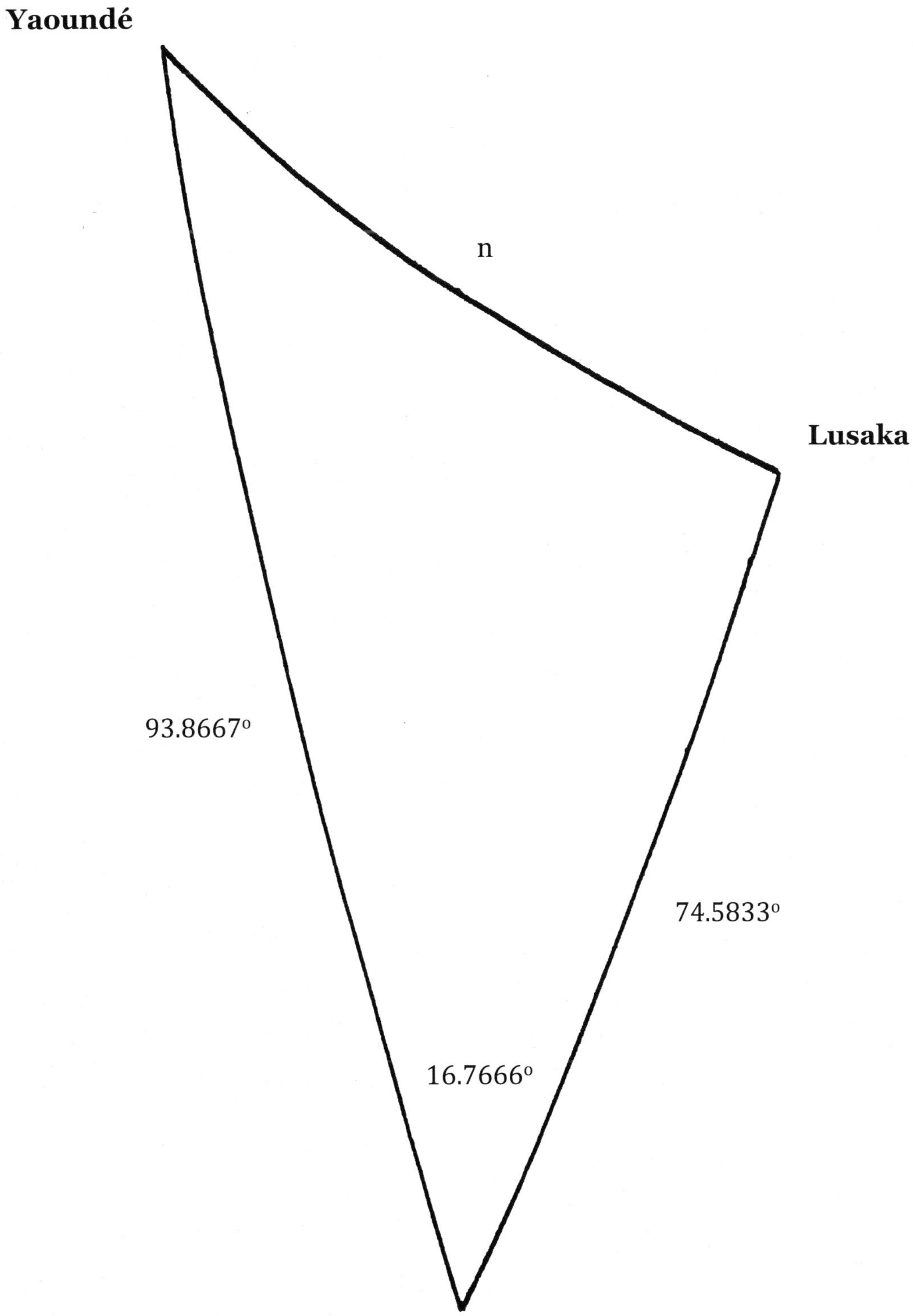
Yaoundé
n
Lusaka
93.8667°
74.5833°
16.7666°
South Pole

Chapter 2

Problem 7

Solution

Praia, Cabo Verde is located at $14^o 55' \text{ N} = \left(14 + {}^{55}/_{60}\right)^o \text{ N} \approx 14.9167^o \text{ N}$,

$23^o 31' \text{ W} = \left(23 + {}^{31}/_{60}\right)^o \text{ W} \approx 23.5167^o \text{ W}$.

N'Djamena, Chad is located at $12^o 6' \text{ N} = \left(12 + {}^{6}/_{60}\right)^o \text{ N} = 12.1000^o \text{ N}$,

$15^o 2' \text{ E} = \left(15 + {}^{2}/_{60}\right)^o \text{ E} \approx 15.0333^o \text{ E}$.

Two sides of the spherical triangle are $90^o - 14.9167^o = 75.0833^o$ and $90^o - 12.1000^o = 77.9000^o$, respectively. The angle at the North Pole is $23.5167^o + 15.0333^o = 38.5500^o$.

See the illustration on the next page.

It follows that

$\cos \text{n} = \cos(75.0833^o)\cos(77.9000^o) + \sin(75.0833^o)\ \sin(77.9000^o)\cos(38.5500^o)$

$\cos \text{n} \approx 0.792879308617 \ \Rightarrow\ \text{n} \approx \cos^{-1}(0.792879308617) \ \Rightarrow\ \text{n} \approx 37.544592592^o$.

Therefore, the approximate distance between the two capital cities is

$(37.544592592^o)\left({}^{\pi}/_{180^o}\right)(3959 \text{ miles})$

$\approx (0.65527675705)(3959 \text{ miles})$

$\approx\ 2{,}594\ \text{ miles}$.

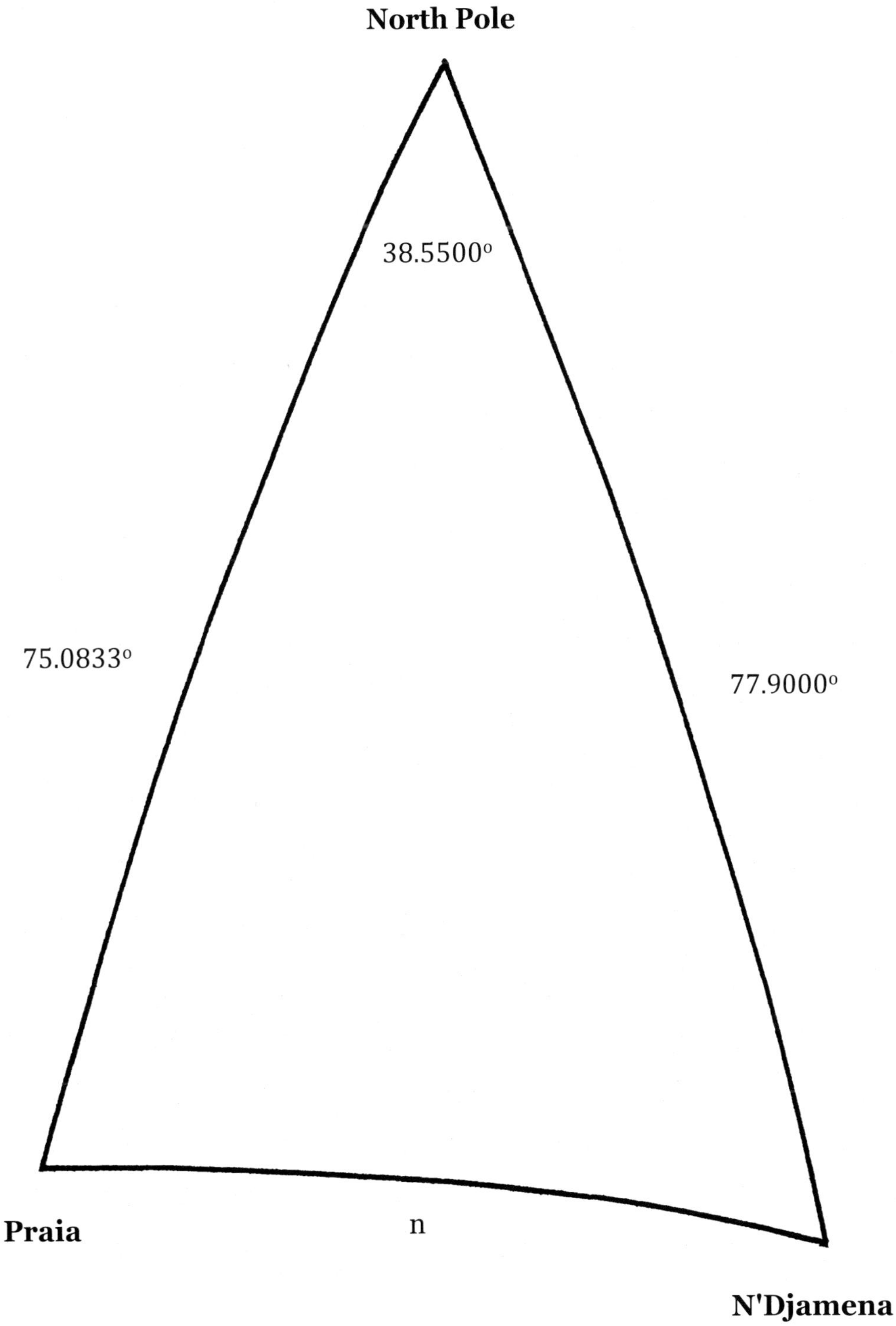
North Pole
38.5500°
75.0833°
77.9000°
Praia
n
N'Djamena

Chapter 2
Problem 8
Solution

Yamoussoukro, Cote d'Ivoire is located at $6^{o}49'\ \mathrm{N} = \left(6 + {}^{49}/_{60}\right)^{o}\ \mathrm{N} \approx 6.8167^{o}\ \mathrm{N}$,
$5^{o}16'\ \mathrm{W} = \left(5 + {}^{16}/_{60}\right)^{o}\ \mathrm{W} \approx 5.2667^{o}\ \mathrm{W}$.

Tripoli, Libya is located at $32^{o}53'\ \mathrm{N} = \left(32 + {}^{53}/_{60}\right)^{o}\ \mathrm{N} \approx 32.8833^{o}\ \mathrm{N}$,
$13^{o}10'\ \mathrm{E} = \left(13 + {}^{10}/_{60}\right)^{o}\ \mathrm{E} \approx 13.1667^{o}\ \mathrm{E}$.

Two sides of the spherical triangle are $90^{o} - 6.8167^{o} = 83.1833^{o}$ and $90^{o} - 32.8833^{o} = 57.1167^{o}$, respectively. The angle at the North Pole is $5.2667^{o} + 13.1667^{o} = 18.4334^{o}$.

See the illustration on the next page.

It follows that

$$\cos \mathrm{n} = \cos(83.1833^{o}) \cos(57.1167^{o}) + \sin(83.1833)\ \sin(57.1167^{o}) \cos(18.4334^{o})$$

$$\cos \mathrm{n} \approx 0.855501001995\ \Rightarrow\ \mathrm{n} \approx \cos^{-1}(0.855501001995)\ \Rightarrow\ \mathrm{n} \approx 31.1848725296^{o}.$$

Therefore, the approximate distance between the two capital cities is

$$(31.1848725296^{o})\left({}^{\pi}/_{180^{o}}\right)(3959\text{ miles})$$

$$\approx (0.544278702456)(3959\text{ miles})$$

$$\approx\ 2{,}155\ \text{ miles.}$$

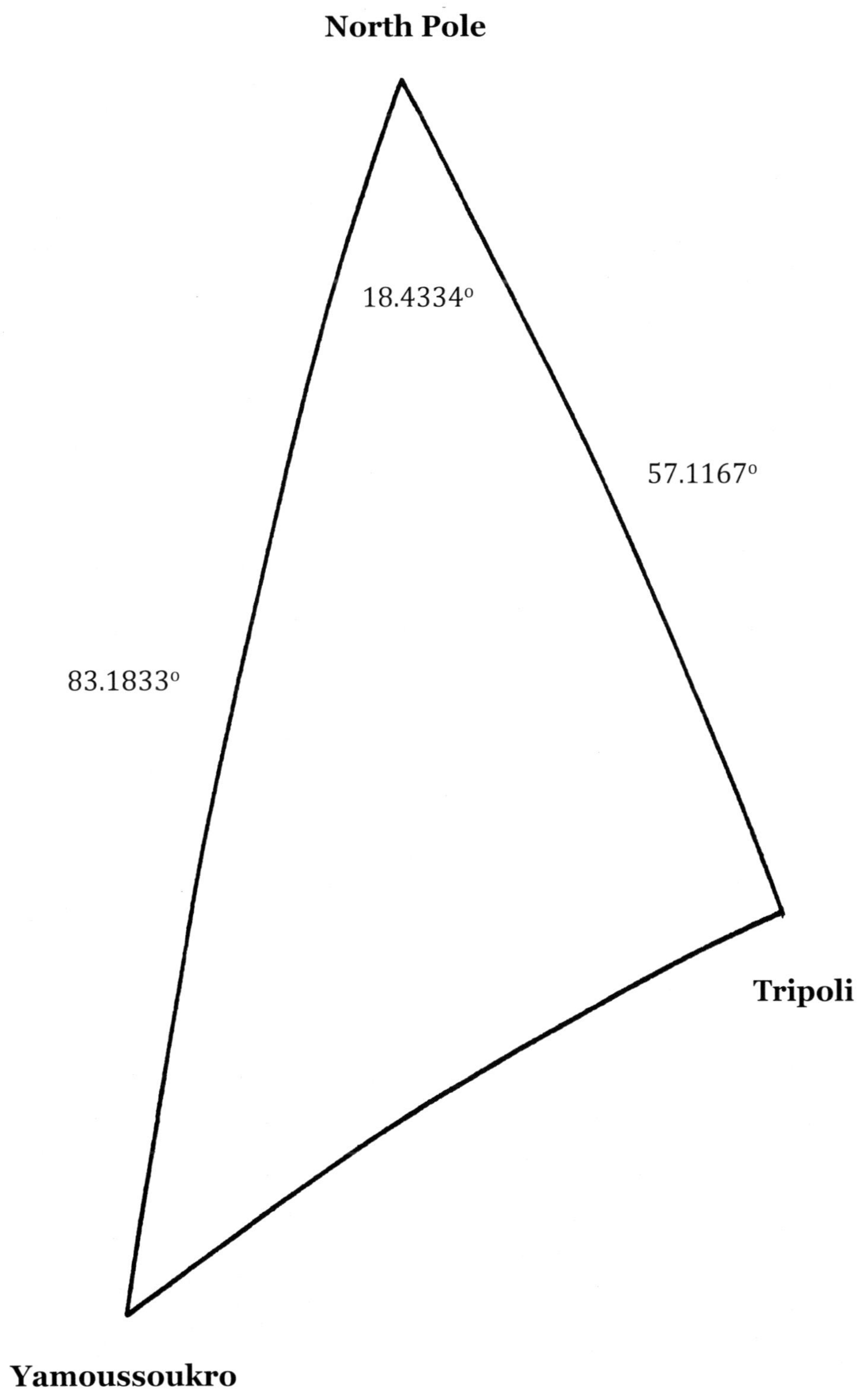
North Pole
18.4334°
57.1167°
83.1833°
Tripoli
Yamoussoukro

Chapter 2
Problem 9
Solution

Kinshasa, Democratic Republic of the Congo is located at

$4^o19'\text{ S} = \left(4 + {}^{19}/_{60}\right)^o \text{ S} \approx 4.3167^o \text{ S}$, $15^o18'\text{ E} = \left(15 + {}^{18}/_{60}\right)^o \text{ E} = 15.3000^o \text{ E}$.

Nairobi, Kenya is located at $1^o17'\text{ S} = \left(1 + {}^{17}/_{60}\right)^o \text{ S} \approx 1.2833^o \text{ S}$,

$36^o49'\text{ E} = \left(36 + {}^{49}/_{60}\right)^o \text{ E} \approx 36.8167^o \text{ E}$.

The sides of the spherical triangle are $90^o - 4.3167^o = 85.6833^o$ and $90^o - 1.2833^o = 88.7167^o$, respectively. The angle at the South Pole is $36.8167^o - 15.3000^o = 21.5167^o$.

See the illustration on the next page.

It follows that

$\cos n = \cos(85.6833^o)\cos(88.7167^o) + \sin(85.6833^o)\ \sin(88.7167^o)\cos(21.5167^o)$

$\cos n \approx 0.929124682637 \ \Rightarrow\ n \approx \cos^{-1}(0.929124682637) \ \Rightarrow\ n \approx 21.7012224269^o$.

Therefore, the approximate distance between the two capital cities is

$(21.7012224269^o)\left({}^{\pi}/_{180^o}\right)(3959 \text{ miles})$

$\approx (0.378757783057)(3959 \text{ miles})$

$\approx\ 1{,}500\ \text{ miles}$.

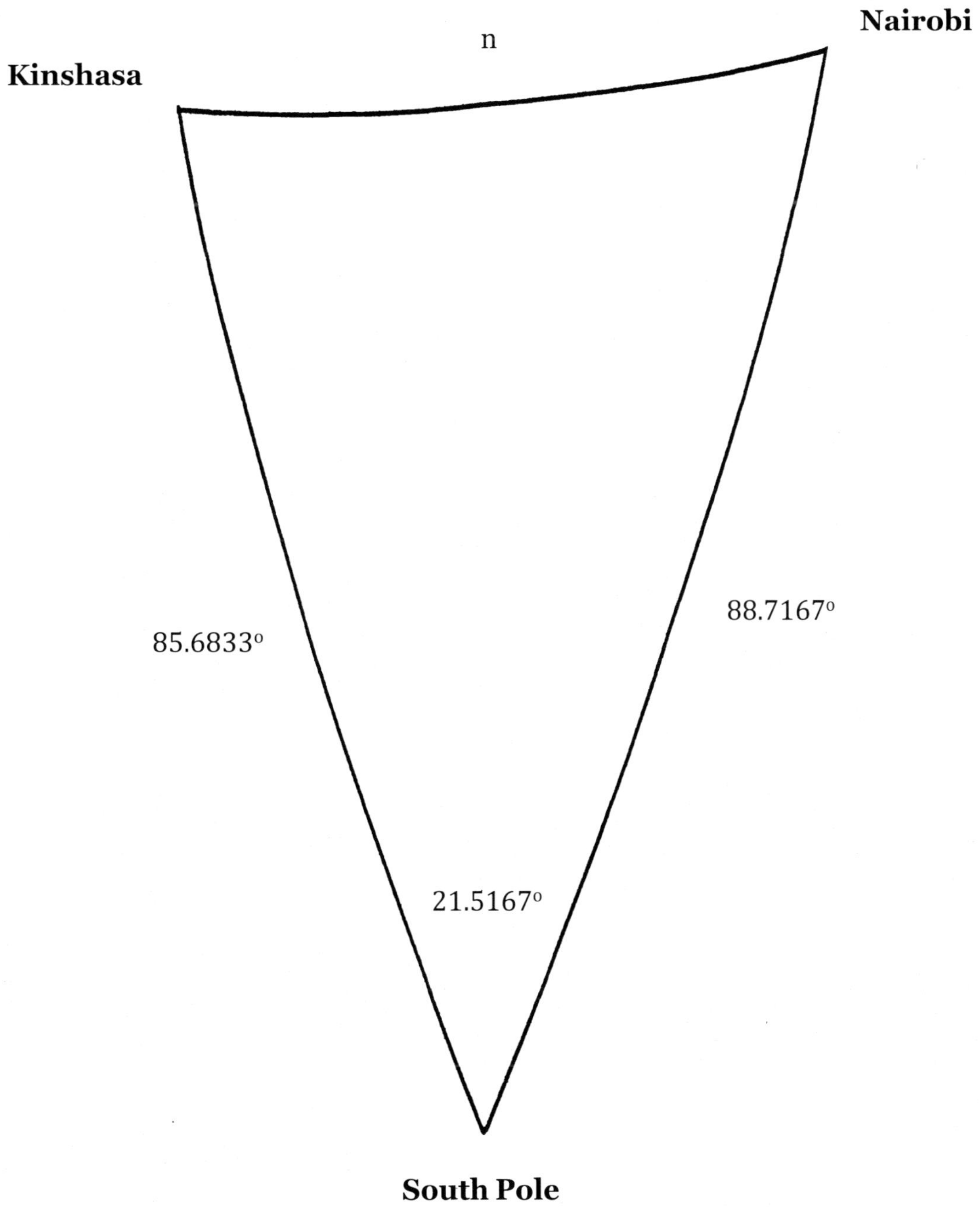
Nairobi
n
Kinshasa
85.6833°
88.7167°
21.5167°
South Pole

Chapter 2
Problem 10
Solution

Banjul, The Gambia is located at $13^o 27' \text{ N} = \left(13 + {}^{27}/_{60}\right)^o \text{ N} = 13.4500^o \text{ N}$,
$16^o 34' \text{ W} = \left(16 + {}^{34}/_{60}\right)^o \text{ W} \approx 16.5667^o \text{ W}$.

São Tomé, São Tomé and Principe is located at $0^o 20' \text{ N} = \left({}^{20}/_{60}\right)^o \text{ N} \approx 0.3333^o \text{ N}$,
$6^o 44' \text{ E} = \left(6 + {}^{44}/_{60}\right)^o \text{ E} \approx 6.7333^o \text{ E}$.

Two sides of the spherical triangle are $90^o - 13.4500^o = 76.5500^o$ and
$90^o - 0.3333^o = 89.6667^o$, respectively. The angle at the North Pole is
$16.5667^o + 6.7333^o = 23.3000^o$.

See the illustration on the next page.

It follows that
$\cos \text{n} = \cos(76.5500^o)\cos(89.6667^o) + \sin(76.5500^o)\ \sin(89.6667^o)\cos(23.3000^o)$

$\cos \text{n} \approx 0.89459433633 \Rightarrow \text{n} \approx \cos^{-1}(0.89459433633) \Rightarrow \text{n} \approx 26.5436289658^o$.

Therefore, the approximate distance between the two capital cities is
$(26.5436289658^o)\left({}^{\pi}/_{180^o}\right)(3959 \text{ miles})$

$\approx (0.46327372088)(3959 \text{ miles})$

$\approx 1{,}834$ miles.

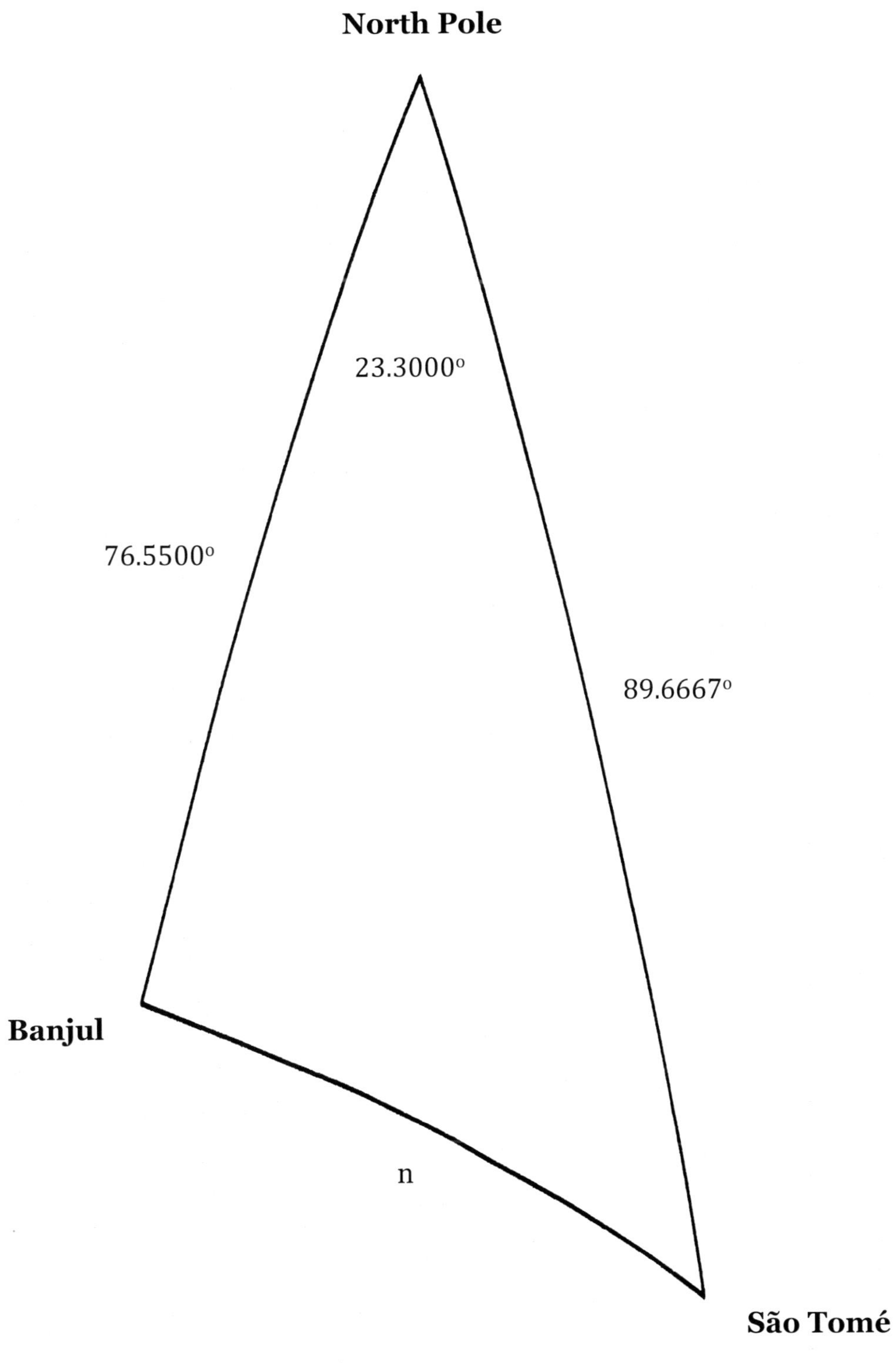
North Pole
23.3000°
76.5500°
89.6667°
Banjul
n
São Tomé

Chapter 2

Problem 11

Solution

Accra, Ghana is located at $5^o33' \text{ N} = \left(5 + {}^{33}/_{60}\right)^o \text{ N} = 5.5500^o \text{ N}$,

$0^o13' \text{ W} = \left({}^{13}/_{60}\right)^o \text{ W} \approx 0.2167^o \text{ W}$.

Moroni, Comoros is located at $11^o42' \text{ S} = \left(11 + {}^{42}/_{60}\right)^o \text{ S} = 11.7000^o \text{ S}$,

$43^o14' \text{ E} = \left(43 + {}^{14}/_{60}\right)^o \text{ E} \approx 43.2333^o \text{ E}$

Two sides of the spherical triangle are $90^o - 5.5500^o = 84.4500^o$ and $90^o + 11.7000^o = 101.7000^o$, respectively. The angle at the North Pole is $0.2167^o + 43.2333^o = 43.4500^o$.

See the illustration on the next page.

It follows that

$\cos \text{n} = \cos(84.4500^o)\cos(101.7000^o) + \sin(84.4500^o)\ \sin(101.7000^o)\cos(43.4500^o)$

$\cos \text{n} \approx 0.687946106747 \ \Rightarrow\ \text{n} \approx \cos^{-1}(0.687946106747) \ \Rightarrow\ \text{n} \approx 46.5322554795^o$.

Therefore, the approximate distance between the two capital cities is

$(46.5322554795^o)\left({}^{\pi}/_{180^o}\right)(3959 \text{ miles})$

$\approx (0.812141066496)(3959 \text{ miles})$

$\approx$ 3,215 miles.

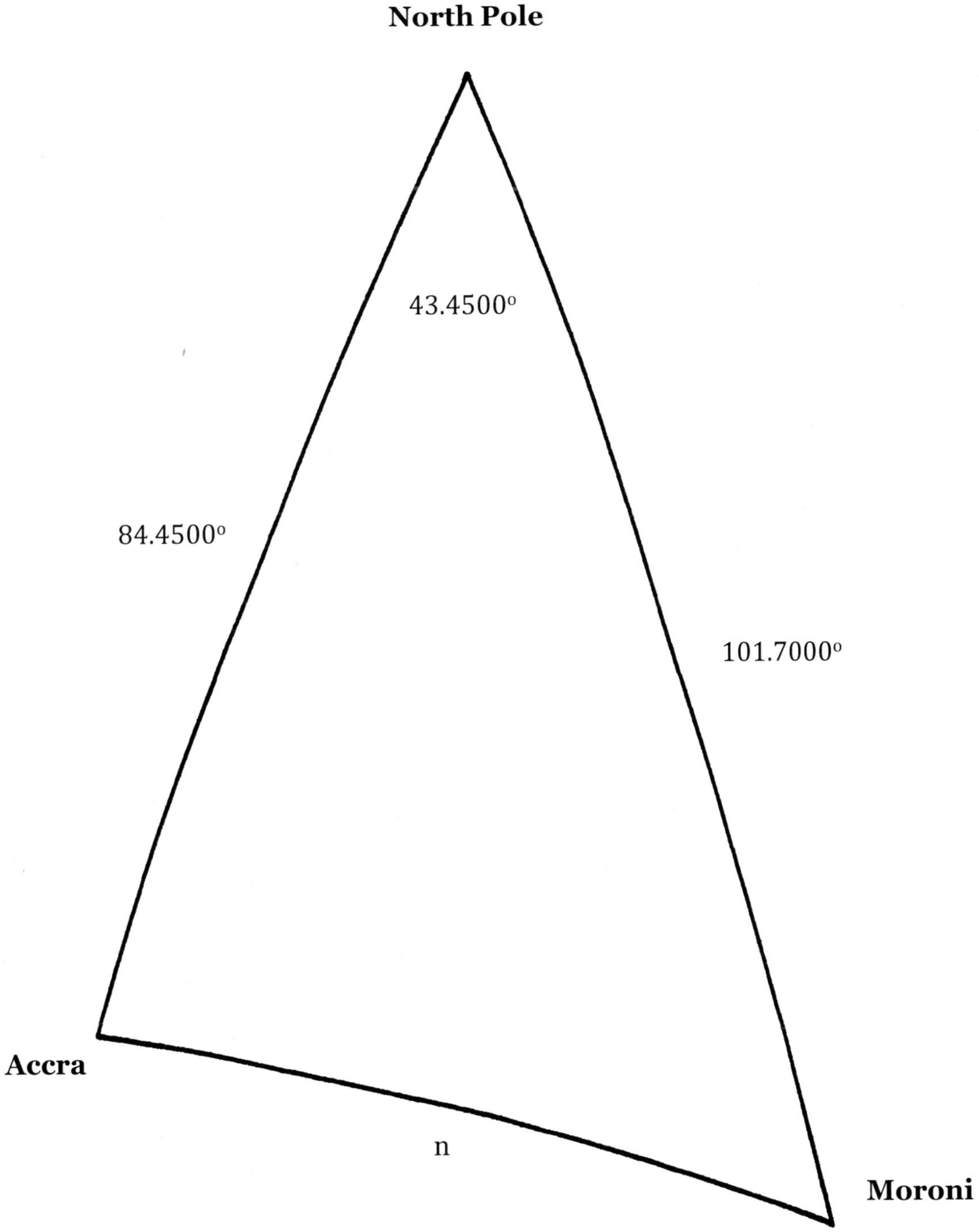
North Pole
43.4500°
84.4500°
101.7000°
Accra
n
Moroni

Chapter 2

Problem 12

Solution

Conakry, Guinea is located at $9^o30'\ \mathrm{N} = \left(9 + {}^{30}/_{60}\right)^o\ \mathrm{N} = 9.5000^o\ \mathrm{N}$,

$13^o42'\ \mathrm{W} = \left(13 + {}^{42}/_{60}\right)^o\ \mathrm{W} = 13.7000^o\ \mathrm{W}$.

Abuja, Nigeria is located at $9^o5'\ \mathrm{N} = \left(9 + {}^{5}/_{60}\right)^o\ \mathrm{N} \approx 9.0833^o\ \mathrm{N}$,

$7^o32'\ \mathrm{E} = \left(7 + {}^{32}/_{60}\right)^o\ \mathrm{E} \approx 7.5333^o\ \mathrm{E}$.

Two sides of the spherical triangle are $90^o - 9.5000^o = 80.5000^o$ and $90^o - 9.0833^o = 80.9167^o$, respectively. The angle at the North Pole is $13.7000^o + 7.5333^o = 21.2333^o$.

See the illustration on the next page.

It follows that

$\cos \mathrm{n} = \cos(80.5000^o)\cos(80.9167^o) + \sin(80.5000^o)\ \sin(80.9167^o)\cos(21.2333^o)$

$\cos \mathrm{n} \approx 0.933857677061 \ \Rightarrow\ \mathrm{n} \approx \cos^{-1}(0.933857677061) \ \Rightarrow\ \mathrm{n} \approx 20.9556284849^o$.

Therefore, the approximate distance between the two capital cities is
$(20.9556284849^o)\left({}^{\pi}/_{180^o}\right)(3959 \text{ miles})$

$\approx (0.365744713886)(3959 \text{ miles})$

$\approx\ 1{,}448\ \text{ miles}$.

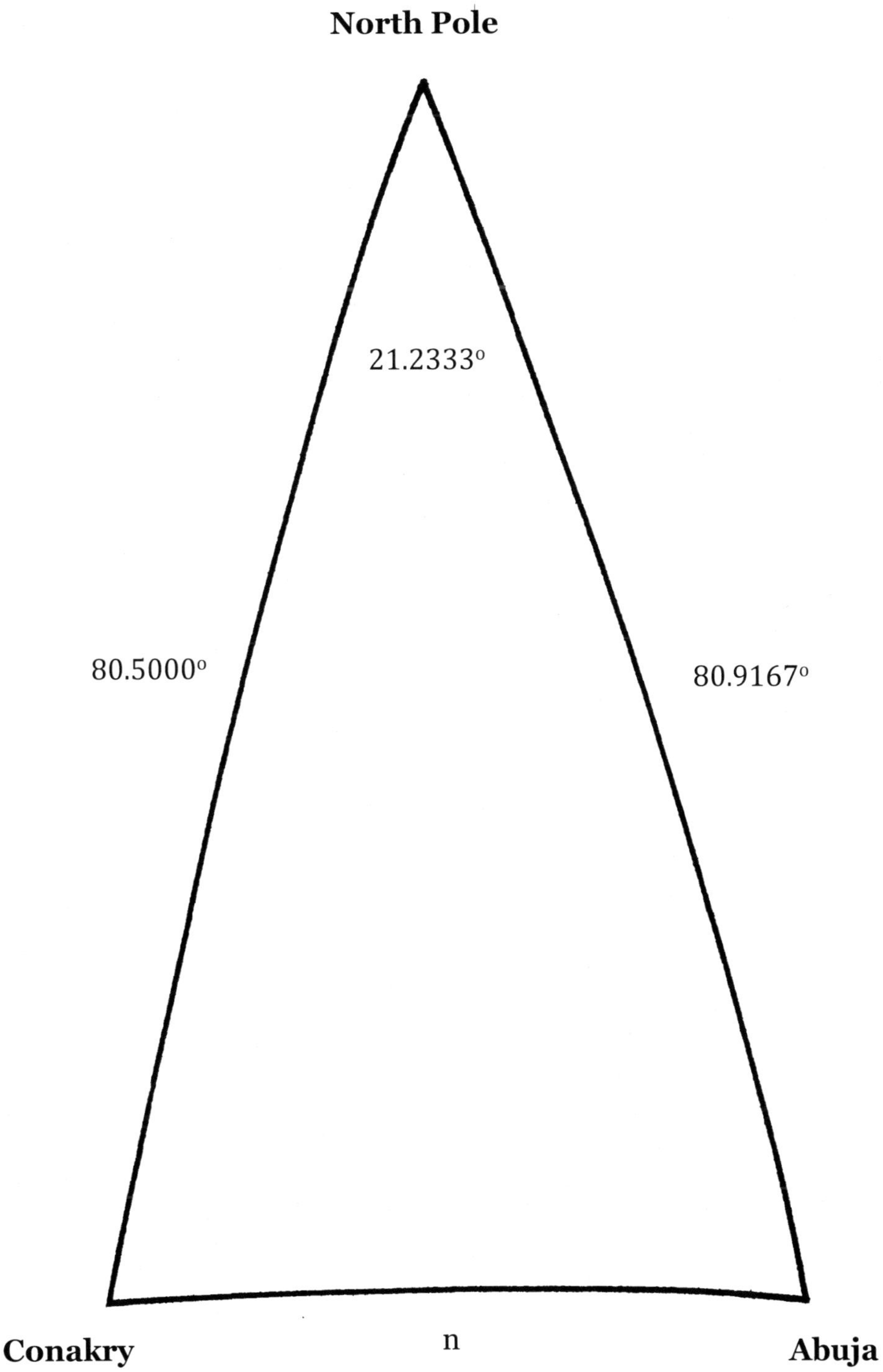
North Pole
21.2333°
80.5000°
80.9167°
Conakry
n
Abuja

Chapter 2
Problem 13
Solution

Maseru, Lesotho is located at $29^o19' \text{ S} = \left(29 + {}^{19}/_{60}\right)^o \text{ S} \approx 29.3167^o \text{ S}$,
$27^o29' \text{ E} = \left(27 + {}^{29}/_{60}\right)^o \text{ E} \approx 27.4833^o \text{ E}$.

Lilongwe, Malawi is located at $13^o58' \text{ S} = \left(13 + {}^{58}/_{60}\right)^o \text{ S} \approx 13.9667^o \text{ S}$,
$33^o47' \text{ E} = \left(33 + {}^{47}/_{60}\right)^o \text{ E} \approx 33.7833^o \text{ E}$.

The sides of the spherical triangle are $90^o - 29.3167^o = 60.6833^o$ and $90^o - 13.9667^o = 76.0333^o$, respectively. The angle at the South Pole is $33.7833^o - 27.4833^o = 6.3000^o$.

See the illustration on the next page.

It follows that

$\cos n = \cos(60.6833^o)\cos(76.0333^o) + \sin(60.6833^o)\ \sin(76.0333^o)\cos(6.3000^o)$

$\cos n \approx 0.959216846415 \ \Rightarrow\ n \approx \cos^{-1}(0.959216846415) \ \Rightarrow\ n \approx 16.4196987834^o$.

Therefore, the approximate distance between the two capital cities is
$(16.4196987834^o)\left({}^{\pi}/_{180^o}\right)(3959 \text{ miles})$

$\approx (0.286577805955)(3959 \text{ miles})$

$\approx\ 1{,}135\ \text{ miles}$.

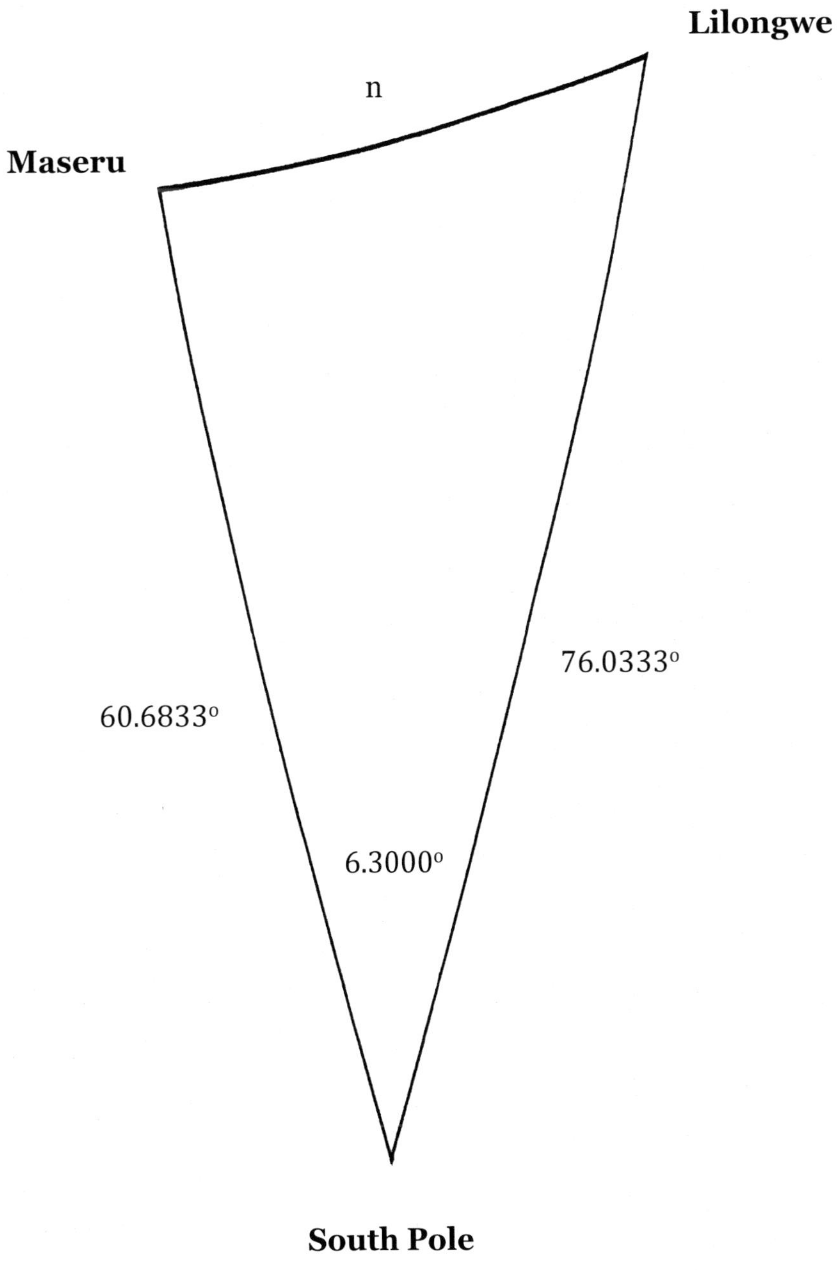
Lilongwe
n
Maseru
76.0333°
60.6833°
6.3000°
South Pole

Chapter 2
Problem 14
Solution

Monrovia, Liberia is located at $6^o18'\text{ N} = \left(6 + {}^{18}/_{60}\right)^o \text{ N} = 6.3000^o \text{ N}$,
$10^o48'\text{ W} = \left(10 + {}^{48}/_{60}\right)^o \text{ W} = 10.8000^o \text{ W}$.

Khartoum, Sudan is located at $15^o36'\text{ N} = \left(15 + {}^{36}/_{60}\right)^o \text{ N} = 15.6000^o \text{ N}$,
$32^o32'\text{ E} = \left(32 + {}^{32}/_{60}\right)^o \text{ E} \approx 32.5333^o \text{ E}$.

Two sides of the spherical triangle are $90^o - 6.3000^o = 83.7000^o$ and $90^o - 15.6000^o = 74.4000^o$, respectively. The angle at the North Pole is $10.8000^o + 32.5333^o = 43.3333^o$.

See the illustration on the next page.

It follows that

$$\cos n = \cos(83.7000^o)\cos(74.4000^o) + \sin(83.7000^o)\sin(74.4000^o)\cos(43.3333^o)$$

$$\cos n \approx 0.72585834893 \Rightarrow n \approx \cos^{-1}(0.72585834893) \Rightarrow n \approx 43.4597007479^o.$$

Therefore, the approximate distance between the two capital cities is
$(43.4597007479^o)\left({}^{\pi}/_{180^o}\right)(3959 \text{ miles})$

$\approx (0.758514869982)(3959 \text{ miles})$

$\approx 3{,}003$ miles.

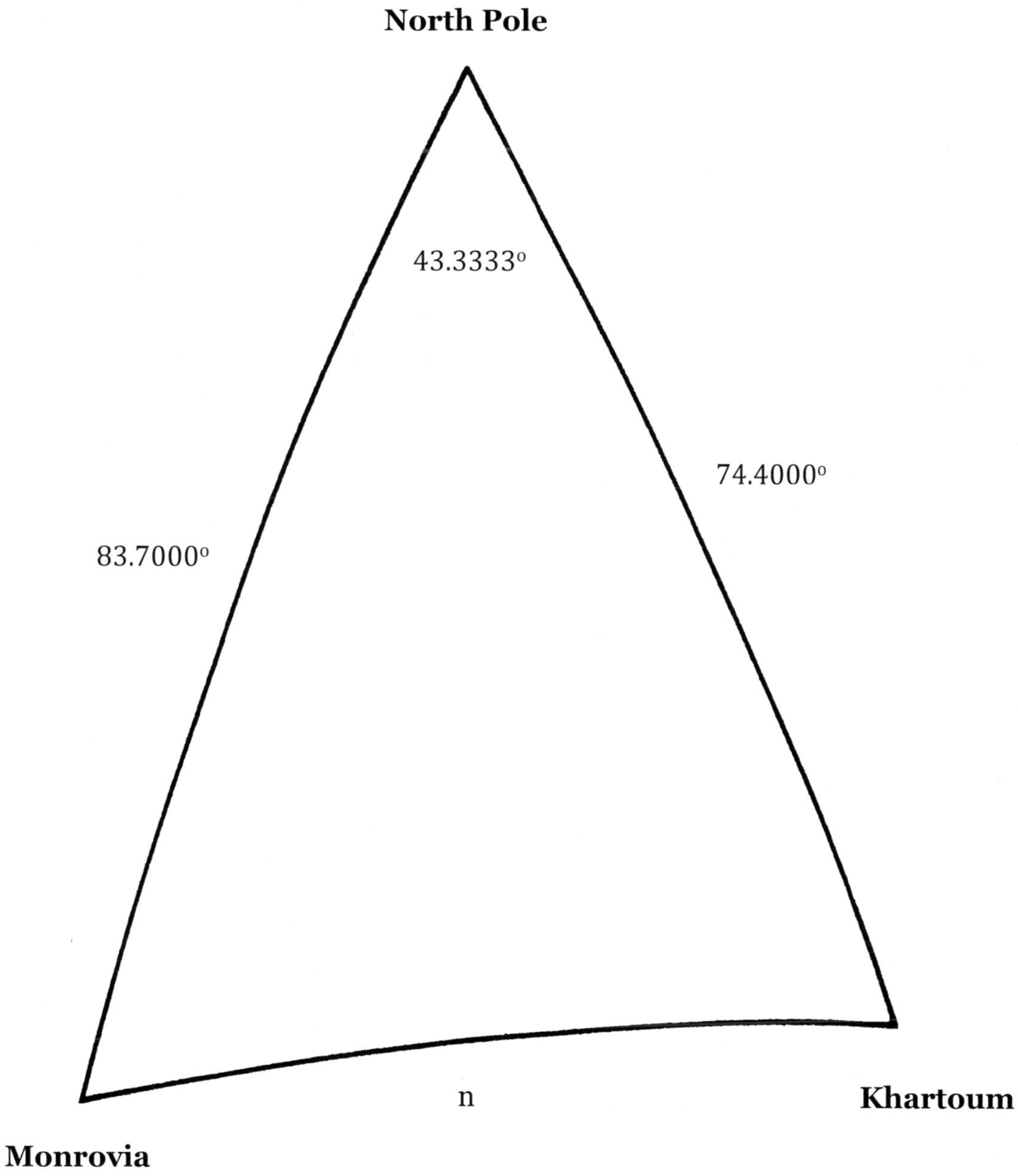
North Pole
43.3333°
74.4000°
83.7000°
n
Khartoum
Monrovia

Chapter 2
Problem 15
Solution

Bamako, Mali is located at $12^o 39' \text{ N} = \left(12 + {}^{39}/_{60}\right)^o \text{ N} = 12.6500^o \text{ N}$, $8^o 0' \text{ W} = 8^o \text{ W}$.

Bangui, Central African Republic is located at

$4^o 22' \text{ N} = \left(4 + {}^{22}/_{60}\right)^o \text{ N} \approx 4.3667^o \text{ N}$, $18^o 35' \text{ E} = \left(18 + {}^{35}/_{60}\right)^o \text{ E} \approx 18.5833^o \text{ E}$.

Two sides of the spherical triangle are $90^o - 12.6500^o = 77.3500^o$ and $90^o - 4.3667^o = 85.6333^o$, respectively. The angle at the North Pole is $8^o + 18.5833^o = 26.5833^o$.

See the illustration on the next page.

It follows that

$$\cos n = \cos(77.3500^o)\cos(85.6333^o) + \sin(77.3500^o)\ \sin(85.6333^o)\cos(26.5833^o)$$

$$\cos n \approx 0.886718084381 \ \Rightarrow\ n \approx \cos^{-1}(0.886718084381) \ \Rightarrow\ n \approx 27.536303455^o.$$

Therefore, the approximate distance between the two capital cities is

$$(27.536303455^o)\left({}^{\pi}/_{180^o}\right)(3959 \text{ miles})$$

$$\approx (0.480599159118)(3959 \text{ miles})$$

$$\approx 1{,}903 \text{ miles.}$$

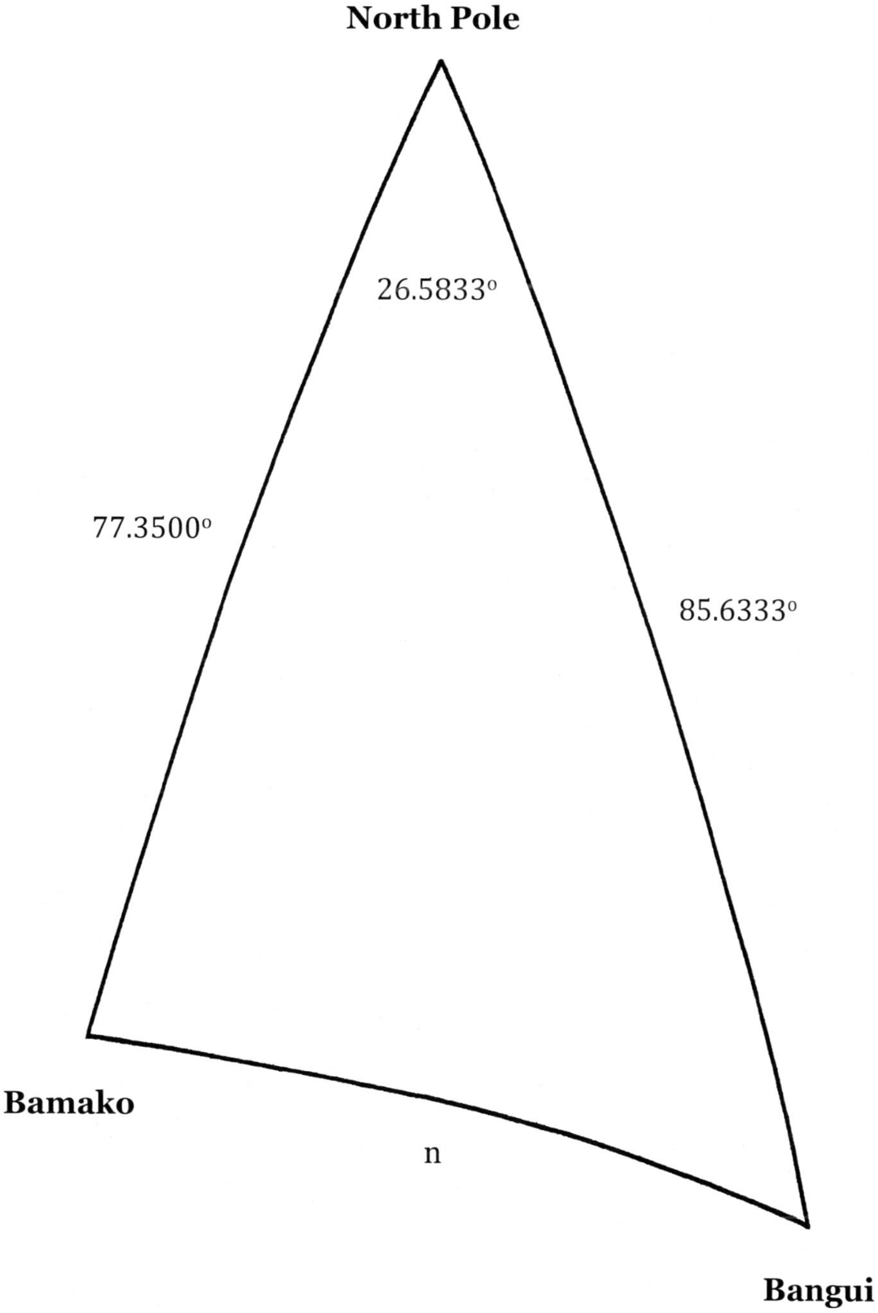
North Pole
26.5833°
77.3500°
85.6333°
Bamako
n
Bangui

Chapter 2
Problem 16
Solution

Nouakchott, Mauritania is located at $18^o4'$ N $= \left(18 + {}^{4}/_{60}\right)^o$ N $\approx 18.0667^o$ N ,

$15^o58'$ W $= \left(15 + {}^{58}/_{60}\right)^o$ W $\approx 15.9667^o$ W .

Victoria, Seychelles is located at $4^o37'$ S $= \left(4 + {}^{37}/_{60}\right)^o$ S $\approx 4.6167^o$ S ,

$55^o27'$ E $= \left(55 + {}^{27}/_{60}\right)^o$ E $= 55.45^o$ E .

Two sides of the spherical triangle are $90^o - 18.0667^o = 71.9333^o$ and $90^o + 4.6167^o = 94.6167^o$, respectively. The angle at the North Pole is $15.9667^o + 55.45^o = 71.4167^o$.

See the illustration on the next page.

It follows that

$\cos n = \cos(71.9333^o)\cos(94.6167^o) + \sin(71.9333^o)\ \sin(94.6167^o)\cos(71.4167^o)$

$\cos n \approx 0.277026036773 \Rightarrow n \approx \cos^{-1}(0.277026036773) \Rightarrow n \approx 73.9172108151^o$.

Therefore, the approximate distance between the two capital cities is

$(73.9172108151^o)\left({}^{\pi}/_{180^o}\right)(3959 \text{ miles})$

$\approx (1.29009870262)(3959 \text{ miles})$

$\approx$ 5,108 miles.

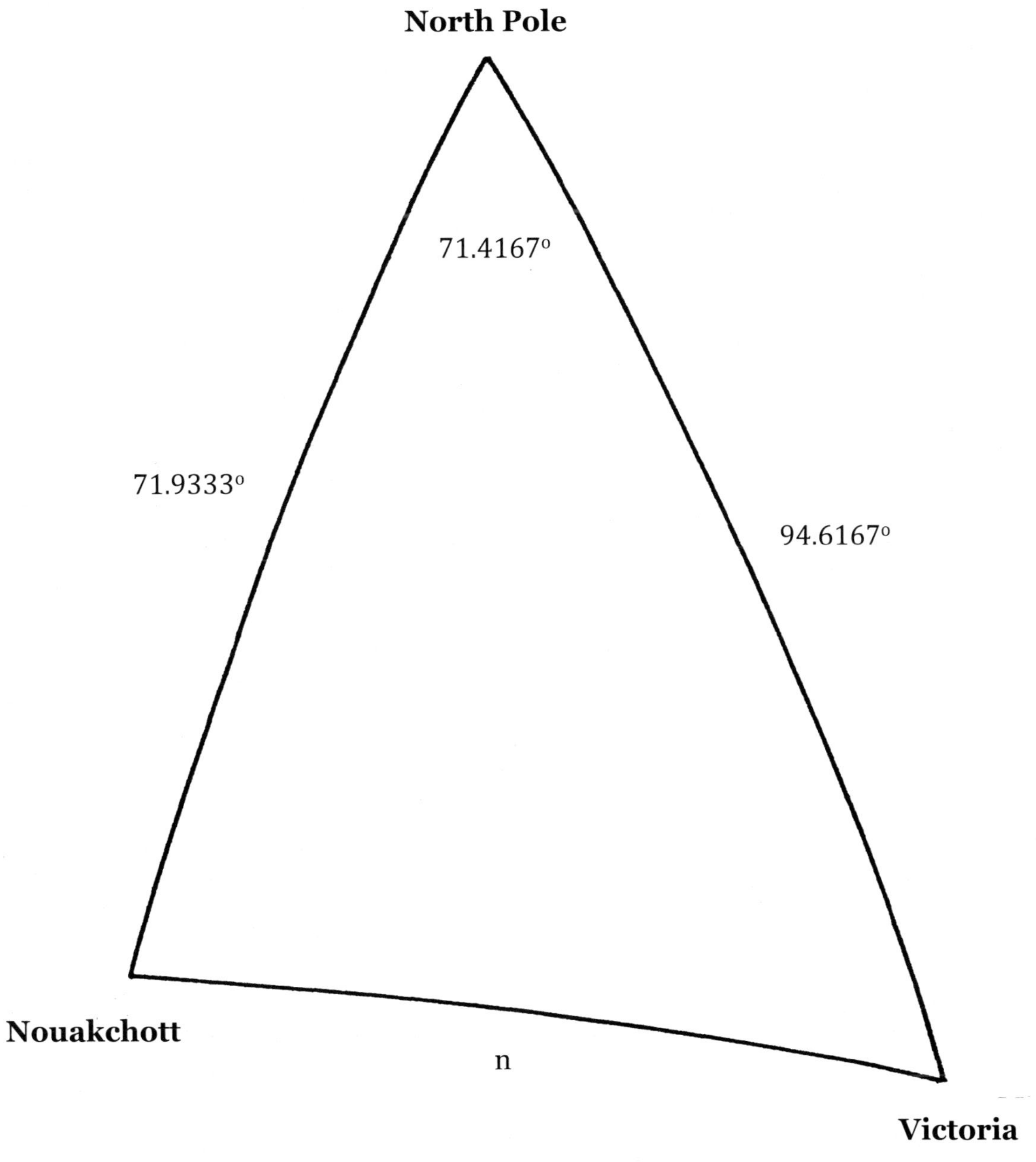
North Pole
71.4167°
71.9333°
94.6167°
Nouakchott
n
Victoria

Chapter 2

Problem 17

Solution

Rabat, Morocco is located at $34^{o}1' \text{ N} = \left(34 + {}^{1}/_{60}\right)^{o} \text{ N} \approx 34.0167^{o} \text{ N}$,

$6^{o}49' \text{ W} = \left(6 + {}^{49}/_{60}\right)^{o} \text{ W} \approx 6.8167^{o} \text{ W}$.

Windhoek, Namibia is located at $22^{o}34' \text{ S} = \left(22 + {}^{34}/_{60}\right)^{o} \text{ S} \approx 22.5667^{o} \text{ S}$,

$17^{o}5' \text{ E} = \left(17 + {}^{5}/_{60}\right)^{o} \text{ E} \approx 17.0833^{o} \text{ E}$.

Two sides of the spherical triangle are $90^{o} - 34.0167^{o} = 55.9833^{o}$ and $90^{o} + 22.5667^{o} = 112.5667^{o}$, respectively. The angle at the North Pole is $6.8167^{o} + 17.0833^{o} = 23.9000^{o}$.

See the illustration on the next page.

It follows that

$$\cos \text{n} = \cos(55.9833^{o})\cos(112.5667^{o}) + \sin(55.9833^{o})\ \sin(112.5667^{o})\cos(23.9000^{o})$$

$$\cos \text{n} \approx 0.485091674029 \ \Rightarrow \ \text{n} \approx \cos^{-1}(0.485091674029) \ \Rightarrow \ \text{n} \approx 60.9815211844^{o}.$$

Therefore, the approximate distance between the two capital cities is

$$(60.9815211844^{o})\left({}^{\pi}/_{180^{o}}\right)(3959 \text{ miles})$$

$$\approx (1.06432832754)(3959 \text{ miles})$$

$$\approx 4{,}214 \text{ miles}.$$

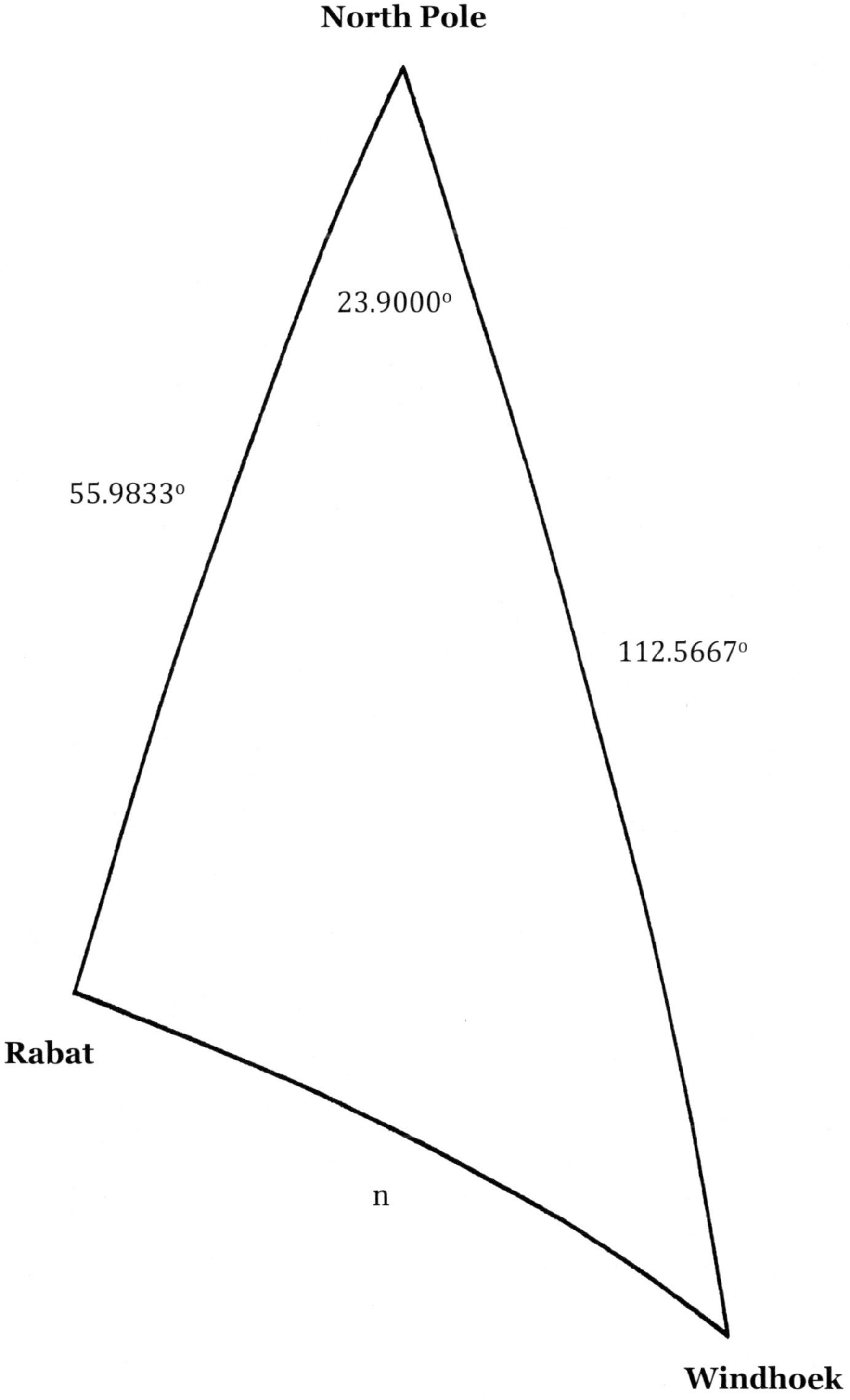
North Pole
23.9000°
55.9833°
112.5667°
Rabat
n
Windhoek

Chapter 2

Problem 18

Solution

Niamey, Niger is located at $13^{o}31' \text{ N} = \left(13 + {}^{31}/_{60}\right)^{o} \text{ N} \approx 13.5167^{o} \text{ N}$,

$2^{o}7' \text{ E} = \left(2 + {}^{7}/_{60}\right)^{o} \text{ E} \approx 2.1167^{o} \text{ E}$.

Kigali, Rwanda is located at $1^{o}57' \text{ S} = \left(1 + {}^{57}/_{60}\right)^{o} \text{ S} = 1.9500^{o} \text{ S}$,

$30^{o}3' \text{ E} = \left(30 + {}^{3}/_{60}\right)^{o} \text{ E} = 30.0500^{o} \text{ E}$.

Two sides of the spherical triangle are $90^{o} - 13.5167^{o} = 76.4833^{o}$ and $90^{o} + 1.9500^{o} = 91.9500^{o}$, respectively. The angle at the North Pole is $30.0500^{o} - 2.1167^{o} = 27.9333^{o}$.

See the illustration on the next page.

It follows that

$\cos \text{n} = \cos(76.4833^{o}) \cos(91.9500^{o}) + \sin(76.4833^{o}) \sin(91.9500^{o}) \cos(27.9333^{o})$

$\cos \text{n} \approx 0.850571745838 \Rightarrow \text{n} \approx \cos^{-1}(0.850571745838) \Rightarrow \text{n} \approx 31.7260898373^{o}$.

Therefore, the approximate distance between the two capital cities is

$(31.7260898373^{o})\left({}^{\pi}/_{180^{o}}\right)(3959 \text{ miles})$

$\approx (0.553724726445)(3959 \text{ miles})$

$\approx 2{,}192 \text{ miles}$.

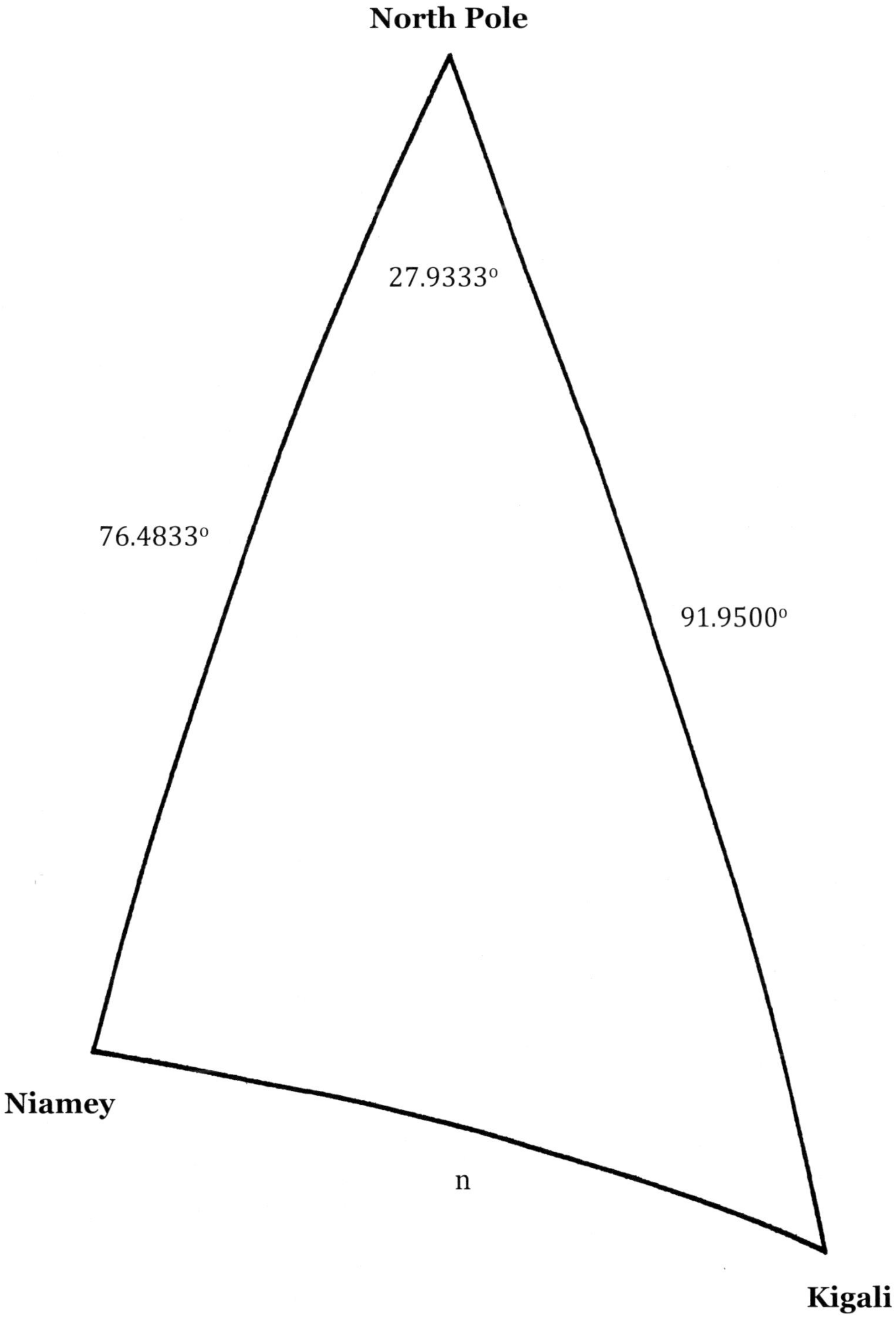
North Pole
27.9333°
76.4833°
91.9500°
Niamey
n
Kigali

Chapter 2

Problem 19

Solution

Brazzaville, Republic of the Congo is located at

$4^o15' \text{ S} = \left(4 + {}^{15}/_{60}\right)^o \text{ S} = 4.2500^o \text{ S}, 15^o17' \text{ E} = \left(15 + {}^{17}/_{60}\right)^o \text{ E} \approx 15.2833^o \text{ E}.$

Djibouti, Djibouti is located at $11^o35' \text{ N} = \left(11 + {}^{35}/_{60}\right)^o \text{ N} \approx 11.5833^o \text{ N}$,

$43^o9' \text{ E} = \left(43 + {}^{9}/_{60}\right)^o \text{ E} = 43.1500^o \text{ E}.$

Two sides of the spherical triangle are $90^o + 4.2500^o = 94.2500^o$ and $90^o - 11.5833^o = 78.4167^o$, respectively. The angle at the North Pole is $43.1500^o - 15.2833^o = 27.8667^o$.

See the illustration on the next page.

It follows that

$\cos \text{n} = \cos(94.2500^o)\cos(78.4167^o) + \sin(94.2500^o)\ \sin(78.4167^o)\cos(27.8667^o)$

$\cos \text{n} \approx 0.848771118449 \quad \Rightarrow \quad \text{n} \approx \cos^{-1}(0.848771118449\ ) \Rightarrow \quad \text{n} \approx 31.9217400319^o.$

Therefore, the approximate distance between the two capital cities is
$(31.9217400319^o)\left({}^{\pi}/_{180^o}\right)(3959 \text{ miles})$

$\approx (0.557139466523)(3959 \text{ miles})$

$\approx$ 2,206 miles.

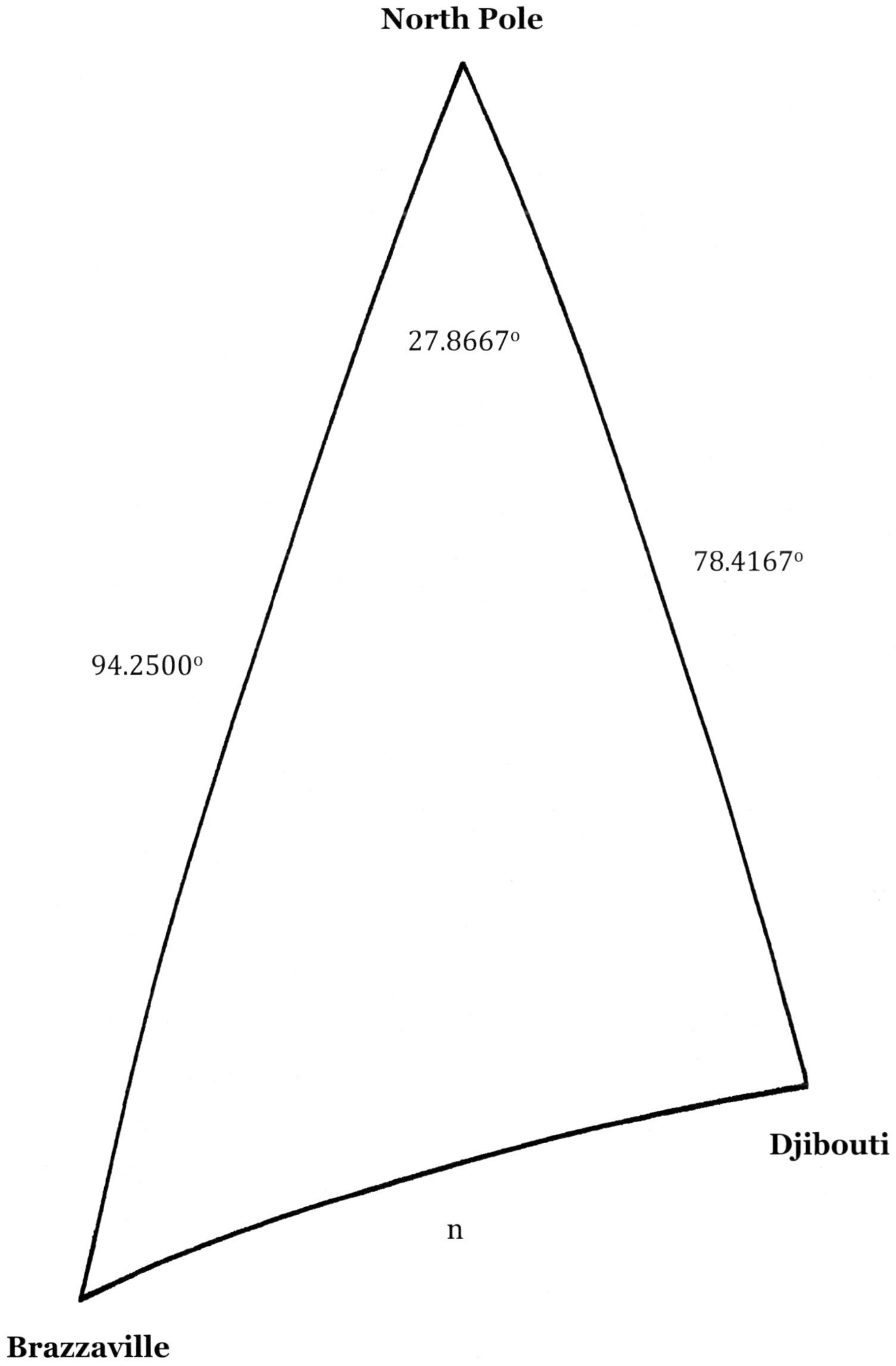
North Pole
27.8667°
78.4167°
94.2500°
Djibouti
n
Brazzaville

Chapter 2

Problem 20

Solution

Dakar, Senegal is located at $14^o44' \text{ N} = \left(14 + {}^{44}/_{60}\right)^o \text{ N} \approx 14.7333^o \text{ N}$,

$17^o38' \text{ W} = \left(17 + {}^{38}/_{60}\right)^o \text{ W} \approx 17.6333^o \text{ W}$.

Cairo, Egypt is located at $30^o3' \text{ N} = \left(30 + {}^{3}/_{60}\right)^o \text{ N} = 30.0500^o \text{ N}$,

$31^o15' \text{ E} = \left(31 + {}^{15}/_{60}\right)^o \text{ E} = 31.2500^o \text{ E}$.

Two sides of the spherical triangle are $90^o - 14.7333^o = 75.2667^o$ and $90^o - 30.0500^o = 59.9500^o$, respectively. The angle at the North Pole is $17.6333^o + 31.2500^o = 48.8833^o$.

See the illustration on the next page.

It follows that

$\cos n = \cos(75.2667^o)\cos(59.9500^o) + \sin(75.2667^o)\ \sin(59.9500^o)\cos(48.8833^o)$

$\cos n \approx 0.677843440925 \;\Rightarrow\; n \approx \cos^{-1}(0.677843440925) \;\Rightarrow\; n \approx 47.3246491072^o$.

Therefore, the approximate distance between the two capital cities is

$(47.3246491072^o)\left({}^{\pi}/_{180^o}\right)(3959 \text{ miles})$

$\approx (0.825970944272)(3959 \text{ miles})$

$\approx \;\; 3{,}270 \;\; \text{miles}.$

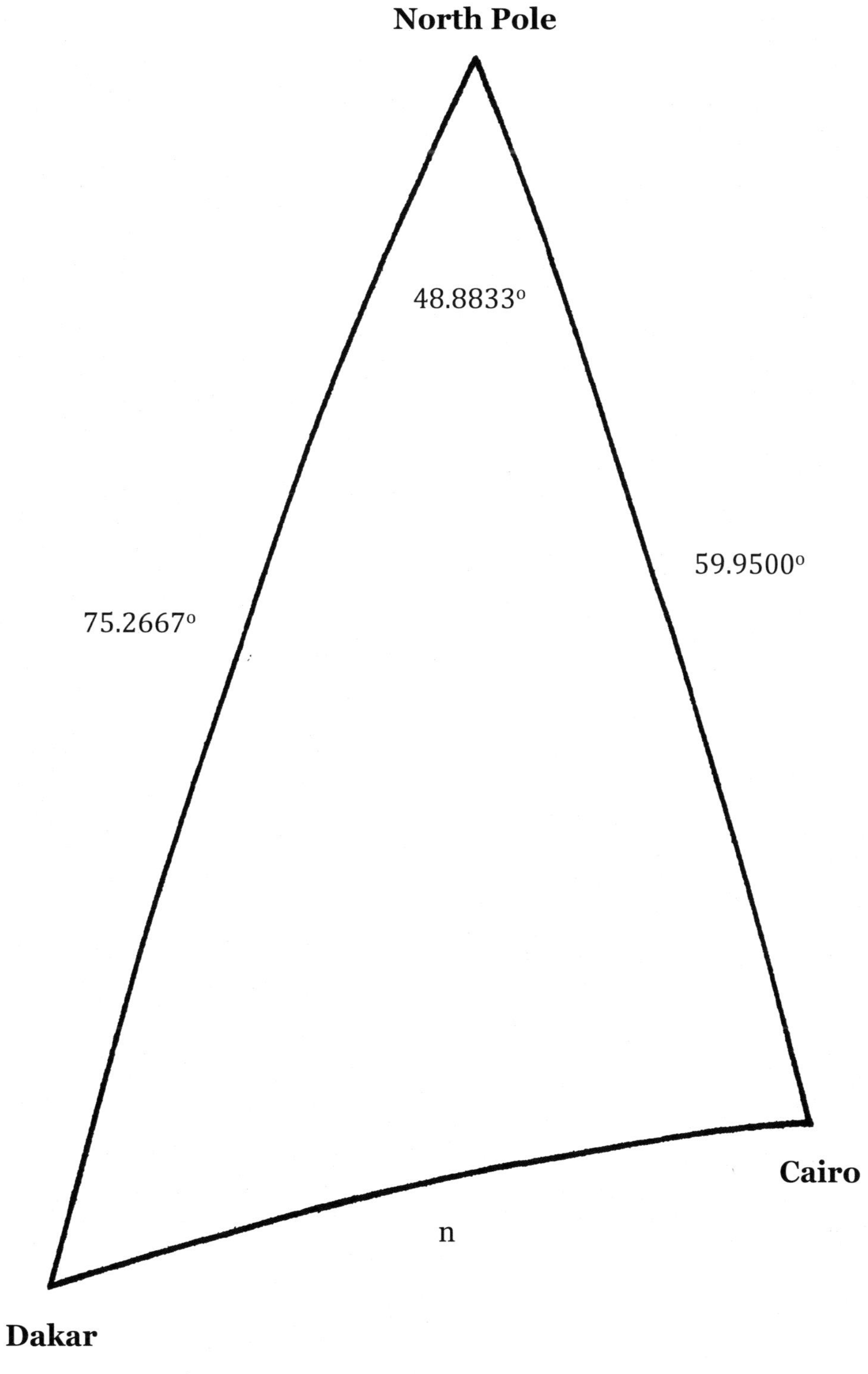
North Pole
48.8833°
59.9500°
75.2667°
Cairo
n
Dakar

Chapter 2
Problem 21
Solution

Freetown, Sierra Leone is located at $8^o 29' \text{ N} = \left(8 + {}^{29}/_{60}\right)^o \text{ N} \approx 8.4833^o \text{ N}$,

$13^o 14' \text{ W} = \left(13 + {}^{14}/_{60}\right)^o \text{ W} \approx 13.2333^o \text{ W}$.

Bujumbura, Burundi is located at $3^o 22' \text{ S} = \left(3 + {}^{22}/_{60}\right)^o \text{ S} \approx 3.3667^o \text{ S}$,

$29^o 21' \text{ E} = \left(29 + {}^{21}/_{60}\right)^o \text{ E} = 29.3500^o \text{ E}$.

Two sides of the spherical triangle are $90^o - 8.4833^o = 81.5167^o$ and $90^o + 3.3667^o = 93.3667^o$, respectively. The angle at the North Pole is $13.2333^o + 29.3500^o = 42.5833^o$.

See the illustration on the next page.

It follows that

$\cos \text{n} = \cos(81.5167^o)\cos(93.3667^o) + \sin(81.5167^o)\ \sin(93.3667^o)\cos(42.5833^o)$

$\cos \text{n} \approx 0.718318276229 \Rightarrow \text{n} \approx \cos^{-1}(0.718318276229\ ) \Rightarrow \text{n} \approx 44.0841918095^o$.

Therefore, the approximate distance between the two capital cities is $(44.0841918095^o)\left({}^{\pi}/_{180^o}\right)(3959 \text{ miles})$

$\approx (0.769414295157)(3959 \text{ miles})$

$\approx\ 3{,}046\ \text{ miles}$.

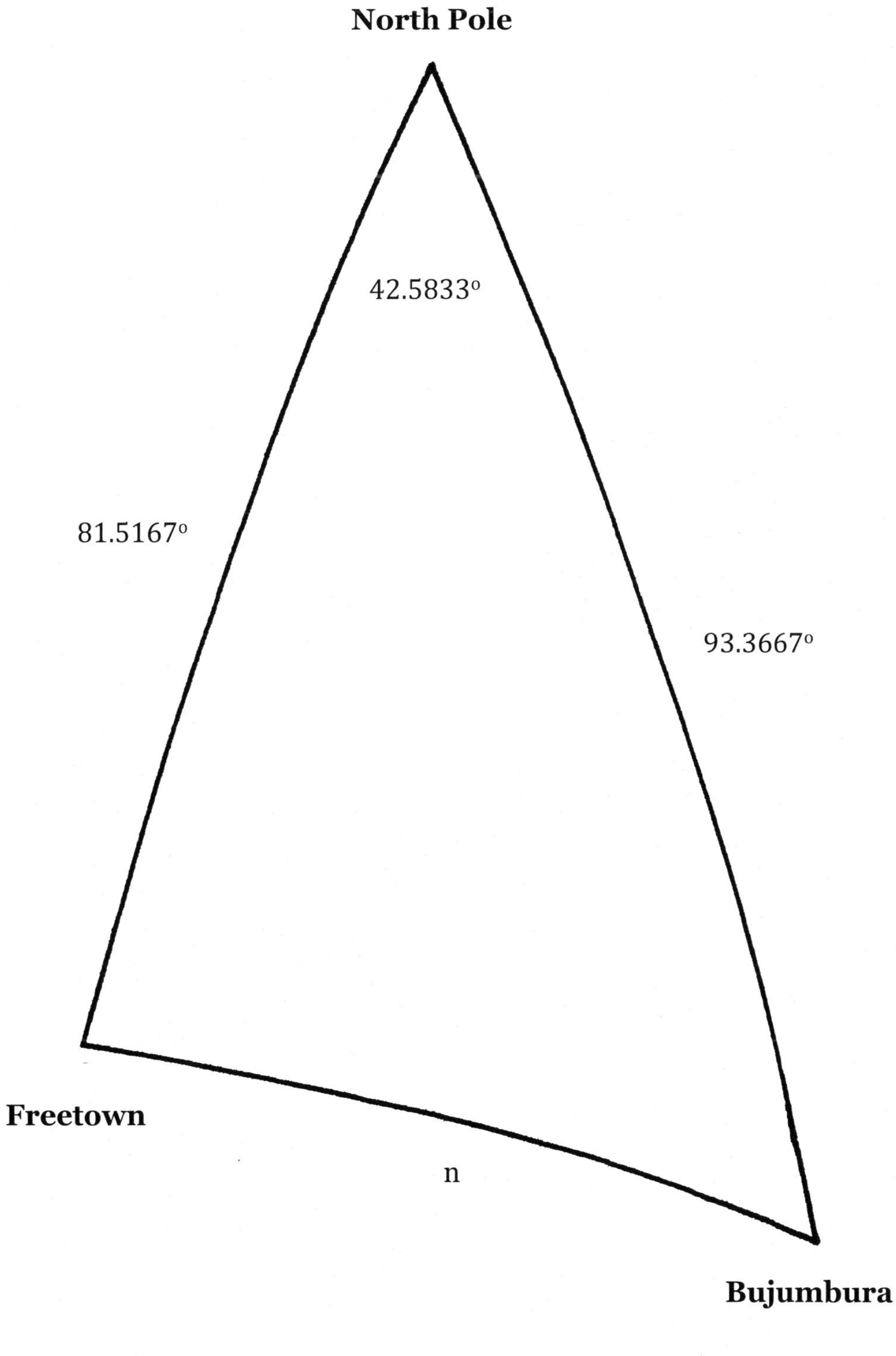
North Pole
42.5833°
81.5167°
93.3667°
Freetown
n
Bujumbura

Chapter 2
Problem 22
Solution

Juba, South Sudan is located at $4^o51'$ N $= \left(4 + {}^{51}/_{60}\right)^o$ N $= 4.8500^o$ N ,

$31^o37'$ E $= \left(31 + {}^{37}/_{60}\right)^o$ E $\approx 31.6167^o$ E .

Mogadishu, Somalia is located at $2^o4'$ N $= \left(2 + {}^{4}/_{60}\right)^o$ N $\approx 2.0667^o$ N ,

$45^o20'$ E $= \left(45 + {}^{20}/_{60}\right)^o$ E $\approx 45.3333^o$ E .

Two sides of the spherical triangle are $90^o - 4.8500^o = 85.1500^o$ and $90^o - 2.0667^o = 87.9333^o$, respectively. The angle at the North Pole is $45.3333^o - 31.6167^o = 13.7166^o$.

See the illustration on the next page.

It follows that

$\cos \text{n} = \cos(85.1500^o) \cos(87.9333^o) + \sin(85.1500^o) \sin(87.9333^o) \cos(13.7166^o)$

$\cos \text{n} \approx 0.970421394864 \Rightarrow \text{n} \approx \cos^{-1}(0.970421394864) \Rightarrow \text{n} \approx 13.9702061016^o$.

Therefore, the approximate distance between the two capital cities is

$(13.9702061016^o)\left({}^{\pi}/_{180^o}\right)(3959 \text{ miles})$

$\approx (0.243826093655)(3959 \text{ miles})$

$\approx$ 965 miles.

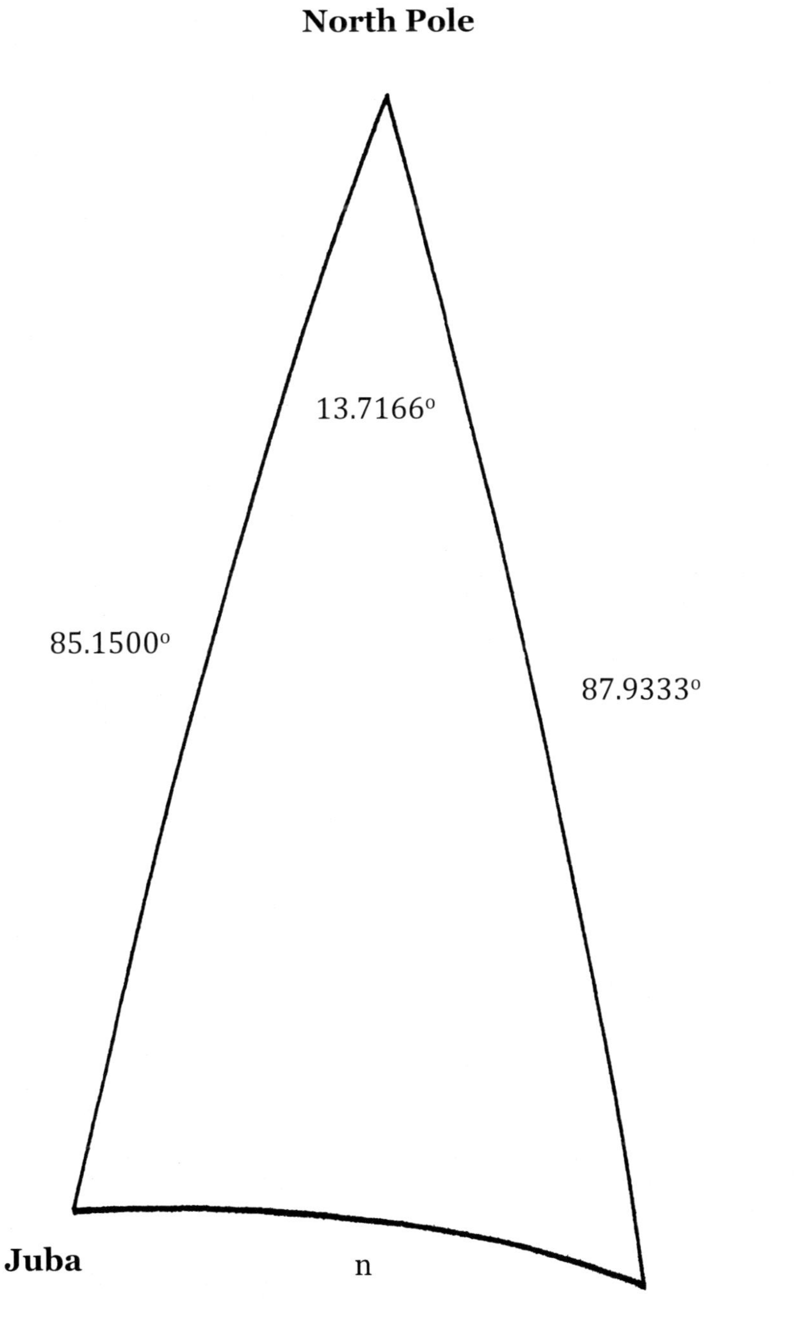

North Pole
13.7166°
85.1500°
87.9333°
Juba
n
Mogadishu

Chapter 2

Problem 23

Solution

Lomé, Togo is located at $6^o7'\ \mathrm{N} = \left(6 + {}^{7}/_{60}\right)^o\ \mathrm{N} \approx 6.1167^o\ \mathrm{N}$,

$1^o13'\ \mathrm{E} = \left(1 + {}^{13}/_{60}\right)^o\ \mathrm{E} \approx 1.2167^o\ \mathrm{E}$.

Maputo, Mozambique is located at $25^o57'\ \mathrm{S} = \left(25 + {}^{57}/_{60}\right)^o\ \mathrm{S} = 25.9500^o\ \mathrm{S}$,

$32^o35'\mathrm{E} = \left(32 + {}^{35}/_{60}\right)^o\ \mathrm{E} \approx 32.5833^o\ \mathrm{E}$.

Two sides of the spherical triangle are $90^o - 6.1167^o = 83.8833^o$ and $90^o + 25.9500^o = 115.9500^o$, respectively. The angle at the North Pole is $32.5833^o - 1.2167^o = 31.3666^o$.

See the illustration on the next page.

It follows that

$$\cos n = \cos(83.8833^o)\cos(115.9500^o) + \sin(83.8833^o)\ \sin(115.9500^o)\cos(31.3666^o)$$

$$\cos n \approx 0.716768073964 \ \Rightarrow\ n \approx \cos^{-1}(0.716768073964) \ \Rightarrow\ n \approx 44.2117127499^o.$$

Therefore, the approximate distance between the two capital cities is

$$(44.2117127499^o)\left({}^{\pi}/_{180^o}\right)(3959\ \text{miles})$$

$$\approx (0.771639955431)(3959\ \text{miles})$$

$$\approx\ 3{,}055\ \ \text{miles.}$$

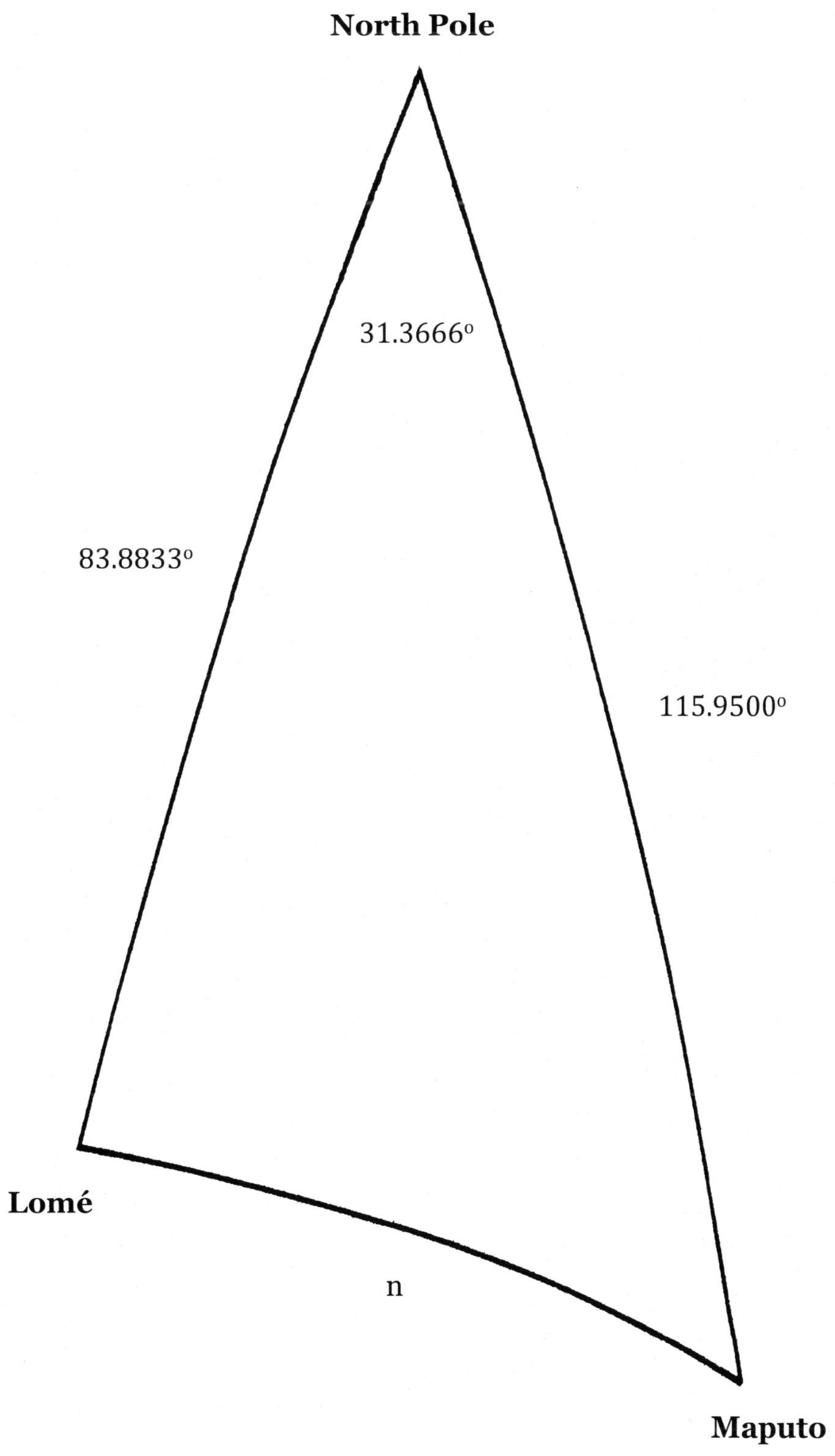
North Pole
31.3666°
83.8833°
115.9500°
Lomé
n
Maputo

Chapter 2
Problem 24
Solution

Tunis, Tunisia is located at $36^{o}48'$ N $= \left(36 + {}^{48}/_{60}\right)^{o}$ N $= 36.8000^{o}$ N ,

$10^{o}11'$ E $= \left(10 + {}^{11}/_{60}\right)^{o}$ E $\approx 10.1833^{o}$ E .

Port Louis, Mauritius is located at $20^{o}9'$ S $= \left(20 + {}^{9}/_{60}\right)^{o}$ S $= 20.1500^{o}$ S ,

$57^{o}29'$ E $= \left(57 + {}^{29}/_{60}\right)^{o}$ E $\approx 57.4833^{o}$ E .

Two sides of the spherical triangle are $90^{o} - 36.8000^{o} = 53.2000^{o}$ and $90^{o} + 20.1500^{o} = 110.1500^{o}$, respectively. The angle at the North Pole is $57.4833^{o} - 10.1833^{o} = 47.3000^{o}$.

See the illustration on the next page.

It follows that

$\cos n = \cos(53.2000^{o})\cos(110.1500^{o}) + \sin(53.2000^{o})\ \sin(110.1500^{o})\cos(47.3000^{o})$

$\cos n \approx 0.3034363126 \Rightarrow n \approx \cos^{-1}(0.3034363126) \Rightarrow n \approx 72.33588656^{o}$.

Therefore, the approximate distance between the two capital cities is

$(72.33588656^{o})\left({}^{\pi}/_{180^{o}}\right)(3959 \text{ miles})$

$\approx (1.262499388)(3959 \text{ miles})$

$\approx\ 4998\ $ miles.

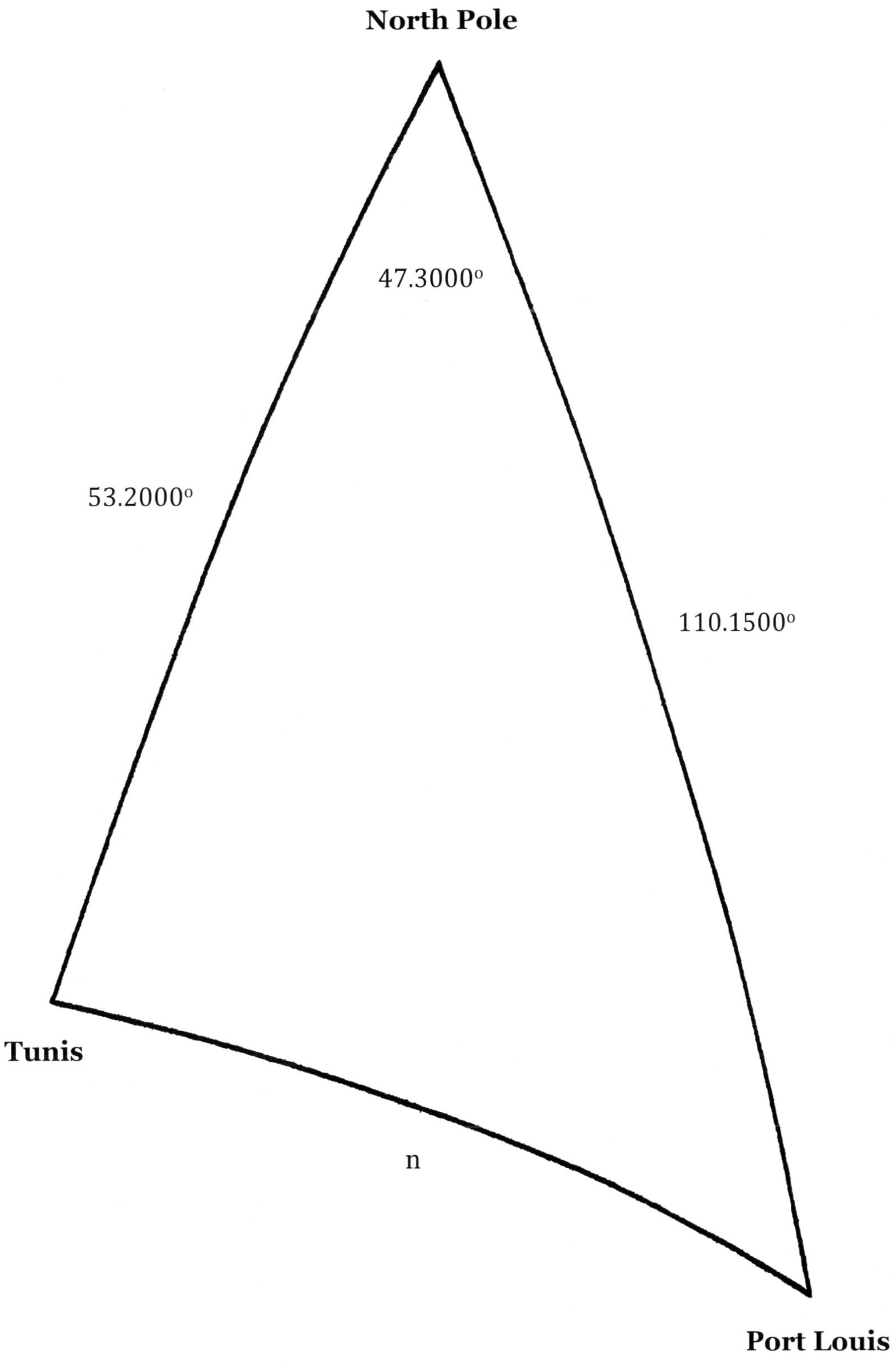
North Pole
47.3000°
53.2000°
110.1500°
Tunis
n
Port Louis

References

Johanson, D. (2017). Approximating Distances Between World Capitals. Pierce, NE: Curve Beneath the Cloud LLC.

Spherical law of cosines. (2017, June 24). Retrieved June 25, 2017, from https://en.wikipedia.org/wiki/Spherical_law_of_cosines

Contact CIA. (2016, April 01). Retrieved June 25, 2017, from https://www.cia.gov/library/publications/resources/the-world-factbook/

Kells, L. M., Kern, W. F., & Bland, J. R. (1942). Spherical trigonometry, with naval and military applications. New York: McGraw-Hill Book Co.

Index of Cities & Countries

M

N

O

P

R

S

T

U

V

W

Y

Z

Made in United States
North Haven, CT
08 November 2026

21928342R00059